P9-AGU-856

Gravity is the most enigmatic of all known basic forces in nature. Yet it controls everything from the motion of ocean tides to the expansion of the entire Universe. Many books use technical jargon and high-powered maths to explain what gravity is all about. In *The lighter side of gravity*, the presentation is beautifully clear and completely nontechnical. Familiar analogies, interesting anecdotes, and numerous illustrations are used throughout to get across subtle effects and difficult points. The coverage is, however, comprehensive and makes no compromise with accuracy. This second edition has been brought completely up to date and expanded to include the discovery of gigantic gravitational lenses in space, the findings of the COBE satellite, the detection of MACHOs, the investigations of the very early Universe, and other new ideas in cosmology. In short, this lucid and stimulating book presents 'the lighter side' of the intriguing phenomena of 'gravity' to the student and general reader.

The lighter side of gravity

# The lighter side of gravity

Second edition

Jayant V. Narlikar

*Inter-University Centre for Astromony and Astrophysics, Pune, India*

Published by the Press Syndicate of the University of Cambridge
The Pitt Building, Trumpington Street, Cambridge CB2 1RP
40 West 20th Street, New York, NY 10011-4211, USA
10 Stamford Road, Oakleigh, Melbourne 3166, Australia

© W. H. Freeman and Company 1982
© Cambridge University Press 1996

First published by W. H. Freeman and Company 1982
Second edition published by Cambridge University Press 1996

Printed in Great Britain at the University Press, Cambridge

*A catalogue record of this book is available from the British Library*

*Library of Congress cataloguing in publication data*

Narlikar, Jayant Vishnu, 1938–
The lighter side of gravity/Jayant V. Narlikar. – 2nd ed.
p. cm.
Includes index.
ISBN 0 521 55009 2. – ISBN 0 521 56565 0 (pbk.)
1. Gravity. I. Title.
QB331.N37 1996
531′.14–dc20 96-12581 CIP

ISBN 0 521 55009 2 hardback
ISBN 0 521 56565 0 paperback

TAG

# *Contents*

# *Preface*

It is often said that modern theoretical physics began with Newton's law of gravitation. There is a good measure of truth in this remark, especially when we take into account the aims and methods of modern physics – to describe and explain the diverse and complex phenomena of nature in terms of a few basic laws.

Gravity is a basic force of the Universe. From the motions of ocean tides to the expansion of the Universe, a wide range of astronomical phenomena are controlled by gravity. Three centuries ago Newton summed up gravity in his simple inverse-square law. Yet, when asked to say why gravity follows such a law, he declined to hazard an opinion, saying 'Non fingo hypotheses' (I do not feign hypotheses). A radically new attempt to understand gravity was made in the early part of this century by Einstein, who saw in it something of deeper significance that linked it to space and time. The modern theoretical physicist is trying to accommodate it within a unified theory of all basic forces. Yet, gravity remains an enigma today.

In this book I have attempted to describe the diversity, pervasiveness, and importance of this enigmatic force. It is fitting that I have focused on astronomical phenomena, because astronomy is the subject that first provided and continues to provide a testing ground for the study of gravity. These phenomena include the motions of planets, comets, and satellites; the structure and evolution of stars; tidal effects on the Earth and in binary star systems; gigantic lenses in space, highly dense objects, such as neutron stars, black holes, and white holes; and the origin and evolution of the Universe itself.

The presentation throughout the book is at a nontechnical level. Although the title is *The lighter side of gravity*, it should not be mistaken for a nonserious presentation. My aim throughout this book has been to emphasize the mutually beneficial interaction between astronomy and gravitational theory, an interaction that has lasted for three centuries and is bound to last into the forseeable future. The adjective 'lighter' is intended to qualify the style of presentation, which is without heavy mathematical formulae or the technical jargon of physics. I hope the lay reader will appreciate the subtle aspects of gravity with the help of the analogies and anecdotes employed instead.

The desire to keep the book compact and more or less along conventional lines has meant the omission of many interesting and fruitful ideas about gravity outside the frameworks of Newton and Einstein. Although the emphasis in this book is on the successes of the ideas of Newton and Einstein and on the daring and speculative applications that these ideas have inspired, I feel that the last word on gravity has not yet been said and that some astronomical phenomena already warrant a fresh input of ideas. Some unconventional ideas, therefore, find a natural place towards the end.

I have enjoyed writing this book, a job that was made easier because of help from so many. My wife Mangala made the initial sketches of the nontechnical figures in the book. She and my parents read the first draft and made valuable suggestions to make it more readable to the lay person.

The first edition of this book, published in 1982, enjoyed a handsome reader response. In the last decade or so there have been further developments in which gravity has played a key role, without (still) giving its secret away. It was therefore a pleasure to respond to Simon Mitton's interest in publishing a revised and updated edition of the book for Cambridge University Press. I have benefited from the advice and comments of Adam Black at CUP on various aspects of this book. I do hope that through this book the reader will share with me the fascination I feel for this strange yet beautiful aspect of nature.

*Jayant V. Narlikar*

# 1

# *Why things move*

## THE RESTLESS UNIVERSE

From ancient Hindu mythology comes this story about the Pole Star: King Uttanapada had two wives. The favourite, Suruchi, was haughty and proud, while the neglected Suniti was gentle and modest. One day Suniti's son Dhruva saw his co-brother Uttama playing on their father's lap. Dhruva also wanted to join him there but was summarily repulsed by Suruchi, who happened to come by. Feeling insulted, the five-year-old Dhruva went in search of a place from where he would not have to move. The wise sages advised him to propitiate the god Vishnu, which Dhruva proceeded to do with a long penance. Finally Vishnu appeared and offered a boon. When Dhruva asked for a place from where he would not have to move, Vishsnu placed him in the location now known as the Pole Star – a position forever fixed.

Unlike other stars and planets, the Pole Star does not rise and set; it is always seen in the same part of the sky. This immovability of the Pole Star has proved to be a useful navigational aid to mariners from ancient to modern times. Yet, a modern-day Dhruva could not be satisfied with the Pole Star as the ultimate position of rest. Let us try to find out why.

The Pole Star does not appear to change its direction in the sky because it happens to lie more or less along the Earth's axis of rotation. As the Earth rotates about its axis, other stars rise over the eastern horizon and set over the western horizon. But as long as the Earth's rotation axis remains unchanged in its direction, the Pole Star will not rise and set but instead will appear fixed,

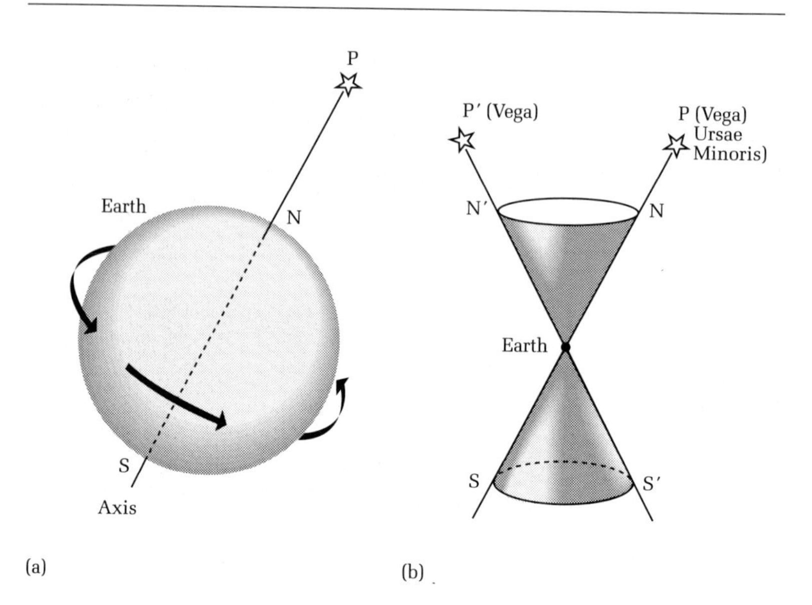


Figure 1-1. (a) The Earth rotates about its north–south axis. The Pole Star $P$ at present lies almost along this axis. (b) The north–south axis is not fixed in space; it *precesses*, or traces the shape of a cone on the sky. Hence, relative to the axis, the Pole Star appears to change direction. Two extreme positions of the axis $NS$ and $N'S'$ are shown as it traces the shape of a cone. At present, the north–south axis points toward the star Alpha Ursae Minoris, commonly known as the Pole Star; after 13 000 years, it will point toward the star Vega, which will therefore replace the Pole Star as the 'fixed' star.

staying always in the same direction. The Earth's rotation axis does, however, change its direction very slowly. Instead of being fixed as in Figure 1-1a, it describes a narrow cone as shown in Figure 1-1b. The time taken for one revolution of the axis along this cone is nearly 26 000 years. No wonder then that, over a human lifetime, or indeed over several centuries, the Pole Star hardly appears to move, whereas in fact it is slowly changing its direction with respect to the Earth's axis.

But this is not the real problem! The Pole Star itself is not fixed in space. Like other stars in our Galaxy, it is moving. Indeed, the Galaxy as a whole (which, as shown in Figure 1-2, is a disk-shaped object with a small bulge in the middle and contains more than 100 billion stars) rotates about its axis with a period of nearly *200*

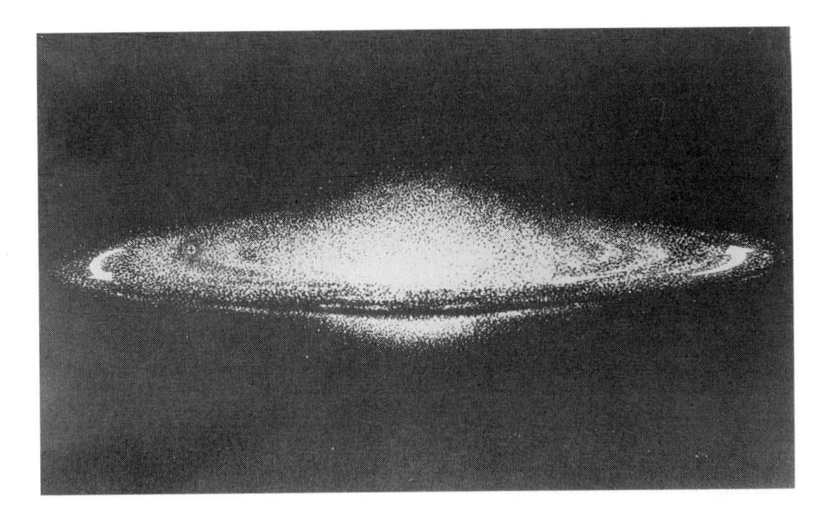

Figure 1-2. A schematic diagram of our Galaxy. The arrows indicate the direction in which the Galaxy rotates. The approximate location of the Pole Star is shown by a circled cross.

*million* years.* So Dhruva cannot really claim to have found a fixed, immovable place!

Indeed, a closer examination shows that mobility rather than rest is the characteristic feature of the Universe. Just as the astronomer discovers examples of motion on the large scale, so the student of microscopic physics finds various examples of motion on the small scale. When we look at a river from a distance, it may appear to us to be at rest. However, when we approach it, we begin to see the steady flow of water. Likewise, on a windless day, we might imagine the air to be at rest. Microscopic physics will tell us, however, that the still air is made up of molecules in *random motion*, and this random motion endows the air with the property of *temperature* (Figure 1-3).

Now, random motion is the very opposite of systematic streamlined motion like that of an army on the march. Air molecules move helter-skelter in arbitrary directions. The faster this random

* Apart from this rotation, our Galaxy takes part in a large-scale motion generally known as the expansion of the Universe. But more of this in Chapter 9.

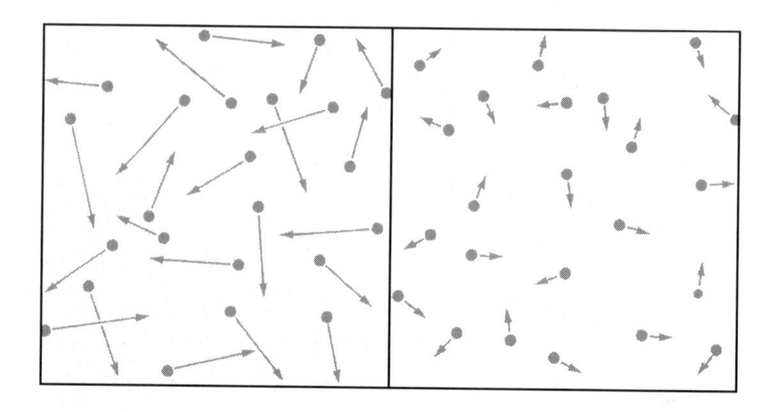

Figure 1-3. Air molecules in random motion. The length of the arrow indicates the speed of the corresponding molecule. In the distribution on the right, the speeds are lower, corresponding to a lower air temperature than in the distribution shown at left.

motion, the higher the air temperature. Going down to the atomic level, we continue to find motion in some form or other in all types of matter. There are lattice vibrations in crystals; there are negatively charged electrons moving freely inside metals; the electrons bound in atoms keep jumping from one orbit to another around the atomic nuclei. Even inside atomic nuclei, things are not at rest! In high-energy particle accelerators like Fermilab near Chicago or CERN near Geneva, atomic physicists are gradually discovering the secrets of the strange world of subatomic particles by bombarding them with one another at high speeds (Figure 1-4). In such a restless Universe, it is going to be a futile exercise to look for rest and immovability. Rather, we should ask the question, 'Why do things move?'

## FROM ARISTOTLE TO GALILEO

This question was posed some twenty-three centuries ago by Aristotle, a Greek philosopher, with the statement that each body has a natural tendency to go to some *preferred position*, and the observed motions in nature show bodies moving *in order to get there.*

Aristotle had been a pupil of Plato, and the tutor of Alexander

Figure 1-4. A section of the high-energy particle accelerator Fermilab. In the tunnel high-speed streams of subatomic particles are generated. (Courtesy of R. R. Wilson, National Accelerator Laboratory.)

the Great. Whereas Alexander's empire crumbled not long after his death, Aristotle's philosophy continued to dominate Europe for several centuries, right through the Middle Ages, and his science came to acquire the authority of the Roman Catholic Church behind it. Today, in the age of modern science, we find Aristotle's approach and ideas strange and difficult to grasp. Yet, when seen against the background of Greece in 350 BC, they reflect a highly systematic attitude.

Aristotle talked about *change* in a system in general terms, and to him motion meant *local* change. Everyday observations present several examples – the motion of stars across the sky, the rising of smoke, the motion of clouds, the fall of rain, tides in the sea, the shooting of arrows, and so on. Aristotle systematized these observations by analysing all the natural examples of motion in terms of combinations of straight motion and circular motion. What is so special about straight lines and circles? As shown in Figure 1-5, these curves are *simple*. Any part of a simple curve can be superimposed on any other part. According to Aristotle's arguments, motion of bodies in nature takes place along such simple curves, or their combinations.

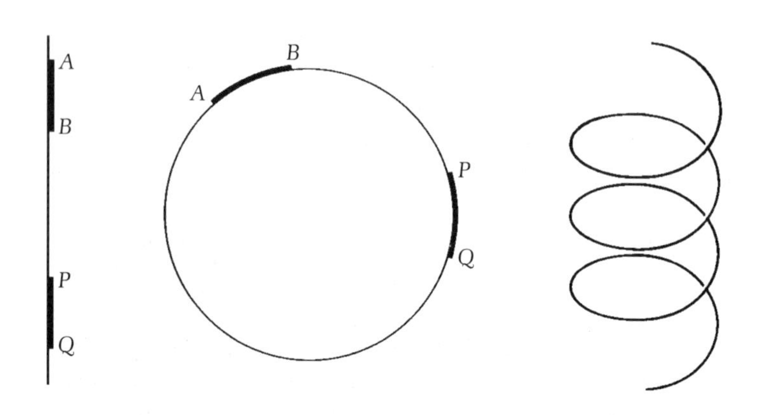


Figure 1-5. The straight line and the circle are simple curves in the sense that any part $AB$ can be superimposed on any other part $PQ$. Galileo pointed out that the helical curve on the right also has this property.

Aristotle, however, distinguished this natural motion from another motion, which he called *violent* motion, that is caused by agents like living beings. Of the examples given above, one, the shooting of an arrow, is by a human being, and Aristotle would argue that this motion of the arrow is not natural but violent. For any such violent motion, there must always be an efficient cause that disturbs the natural tendency to move along simple paths.

One of Aristotle's ideas that took firm root in European culture was the notion of a fixed Earth amidst a revolving cosmos. The so-called *geocentric* theory, which had the Earth as the fixed object in the midst of a moving assembly of the Sun and the planets, had become a religious dogma which received its first major challenge in the work of Copernicus in the sixteenth century. In the *heliocentric* theory of Copernicus it is the Sun that is the fixed centrepiece around which all planets, including the Earth move. We will discuss the Copernican revolution in the following chapter.

It was not until the seventeenth century that a serious challenge to Aristotle's ideas was posed. The man to do so was Galileo Galilei, mathematician and philosopher to the Grand Duke of Florence. Galileo's genius lay not so much in mathematics and philosophy but in clever experimental demonstrations to support his arguments. Galileo's book *Dialogue Concerning the Two Chief World Systems –*

*Ptolemaic and Copernican* is a brilliant demonstration of modern scientific reasoning pitted against the medieval Aristotelian philosophy. Not only did Galileo defend the Copernican system, he also attacked the very foundations of Aristotelian natural philosophy. Because of his demonstrations Galileo may be said to have pioneered the spirit of experimentation in modern science.

It will not be possible here to reproduce, even in a brief summary, all of Galileo's arguments and demonstrations against the Aristotelian system. Let us take two examples relating to violent motion as described by Aristotle: shooting an arrow and pushing a cart.

When an arrow is shot from a bow, why does it move? According to Aristotle, there must be an agent acting on it all the time to cause its motion. First, of course, the human being who shot the arrow supplied the cause. But what next? To keep the arrow flying, the Aristotelians had to argue that the air behind the arrow keeps on pushing it, just as wind pushes clouds in the sky. Galileo's reply to this reasoning was, first, to show that if an arrow is shot sideways, that is, in a direction perpendicular to its length (Figure 1-6), it goes only a short distance. If Aristotle were right, would not air have a greater cross-section of the arrow available to push against, and hence to give a greater speed to, than when the arrow is shot lengthwise in the usual manner?

Another of Aristotle's examples to support his claim that force is needed to sustain motion was that of a pushcart. The cart needs to be pushed in order to move. So long as someone is pushing it, the pushcart goes forward; but when the pushing stops, the cart also comes to a halt. So, argued the Aristotelians, that force is need to keep the movement going. How did Galileo tackle this basic premise?

To understand the key point of difference between Aristotle and Galileo, let us first take the example of a car in motion. A moving car changes its location with time and we specify the speed of the car by dividing the distance covered by it by the time interval. Thus a car with a speed of 60 km/h would cover a distance of 60 kilometres in one hour *if it moved with a constant speed.* In practice the car may not maintain the same speed over one hour. It may slow down (*decelerate*) through braking or go faster (*accelerate*) if the driver presses the gas pedal. But the speedometer indicates the

Figure 1-6. An arrow shot in the direction of its length goes much farther than an arrow shot perpendicular to its length. Galileo cited this experiment to rule out the Aristotelian principle that things move because air pushes them. In this experiment, air has a larger cross-section of the arrow to push against when it is shot sideways than when it is shot lengthwise.

instantaneous speed of the car and the flickering of its needle tells us that the speed is changing.

Aristotle had stated that a constant force would generate a constant speed. To test the correctness of this argument, Galileo constructed a water clock to measure accurate time intervals and then performed the experiment of dropping a heavy body from a great height. If the weight of the body is the force responsible for the fall, according to Aristotle the body should fall through equal heights in equal time intervals. Galileo demonstrated (Figure 1-7) that the body falls through *increasing* heights in equal time intervals. Its speed does *not* remain constant but *increases* in proportion to time.

When a body is dropped, it is initially at rest and then picks up speed. The same thing happens in other examples in daily life. A car is at rest. When it starts moving, what is observed is not a

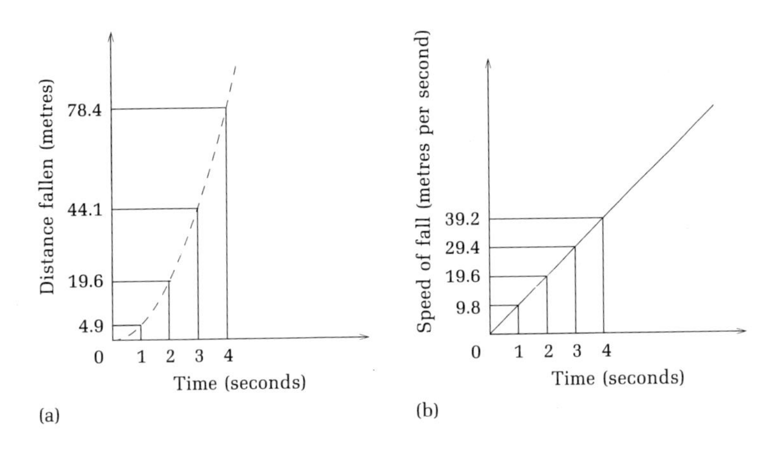


Figure 1-7. Galileo demonstrated by experiments that, as bodies are dropped from a height, they fall through increasing distances in equal time intervals. As shown by the dashed curve (a), equal intercepts of time correspond to increasing intercepts of distance as the body falls. The speed of fall also progressively increases with time, as shown by the solid straight line (b).

constant velocity but a change of velocity. This change of velocity per unit time is known as *acceleration.* If a car is moving at 50 km/h on a smooth surface and the driver further presses the gas pedal, the speed of the car increases – it accelerates. Galileo was able to recognize that the real effect when a force is applied to a body is one of acceleration.

Let us go back to the example of the pushcart. To cause it to move, we need a force. Stop pushing and the cart stops. So was not Aristotle right when he said that force is needed for motion? Of course, the fallacy in the reasoning begins to show up with a little thought. When we begin to push with a certain force, the cart does not immediately acquire a constant speed. Its speed slowly builds up – that is, the cart accelerates (just as the car does when we press the gas pedal). When we withdraw the force, the cart should stop at once, if Aristotle were right. However, it continues to move for a while with diminishing speed before it stops. The cart decelerates because a force has been acting on it all along, opposing its motion. This is the force of *friction.* The pushing force is needed to counteract friction in order to keep the cart moving at constant speed. When the pushing force is withdrawn, the downward change of speed is caused by the force of friction.

## THE LAWS OF MOTION

By arguments and experiments of this kind, Galileo correctly grasped the relationship between force and motion. He realized that force causes a change of the state of motion and that if there is no force acting on a body there will be no change in its speed. Galileo's appreciation of the relationship between the force and the change of motion was only a qualitative one. A quantitative statement of this relationship had to wait for a few decades after Galileo. A precise statement of the laws of motion was given by Isaac Newton, who was born the year that Galileo died (1642). In his book *The Mathematical Principles of Natural Philosophy*, published in 1687, Newton gave a detailed discussion of these three laws of motion.

Before we consider Newton's laws, let us define some concepts of *dynamics*, the subject dealing with the motion of bodies under different forces. We begin by highlighting the distinction between *speed* and *velocity*. We have already mentioned speed. The concept of velocity actually includes two bits of information: how fast the body is moving and in what direction it is moving. The first bit of information signifies speed. Thus, the information that a car is moving at 60 km/h tells us about the speed of the car. To know its velocity, we must know also the direction in which it is going.

Acceleration, as mentioned before, denotes the *rate of change* of velocity. A change in the velocity could occur in two ways: through a change in the speed or a change in the direction. In Figure 1-8, we see a car going around a circular race track with a constant speed of 150 km/h. Although its speed is constant, its direction is changing all the time as it travels in the circle. Therefore, the car is accelerating.

Like velocity, acceleration also has magnitude and direction. What is the acceleration of the car in this case? The answer is usually obtained with the aid of calculus, but it can be stated in a simple form. The acceleration is directed toward the centre of the race track and has a magnitude equal to the square of the speed of the car divided by the radius of the track. If the track has a one-kilometre

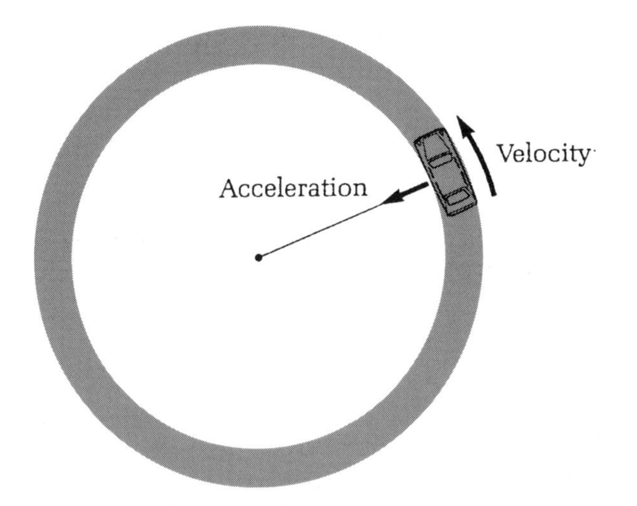


Figure 1-8. A car moving on a circular track has a continuously changing direction. It may have a constant speed, but its velocity changes because of the change of direction. The car has a resulting acceleration toward the centre of the track.

radius, the magnitude of the acceleration is 150 km/h × 150 km/h ÷ 1 km = 22 500 km/h/h.

Another useful concept is that of *angular velocity*. In the example of the race car, when it has made one complete round, we say it has completed a total angle of 360° about the centre. How long does the car take to make one round? We already know that the speed of the car is 150 km/h and the circumference of the track is $2\pi$ kilometres.* Therefore, the time taken to make one round of the track is $2\pi/150$ hours, or about $2\frac{1}{2}$ minutes. Since during this time the car traverses an angle of 360°, its angular velocity has a magnitude of about $360°/2\frac{1}{2}$, or 144° per minute.

Now, to return to the laws of motion, the first law was already known to Galileo. It states that a body will continue to be in a state of rest or of uniform velocity unless some external force acts on it. The type of experiment and reasoning that led Galileo to this law is described in Figure 1-9. Notice the contrast between this law and Aristotle's concept of motion. Aristotle required a constant force to

* The circumference of a circle is $2\pi$ times its radius. The constant $\pi$ = 3.141 596..., and it is often approximated by the fraction 22/7.

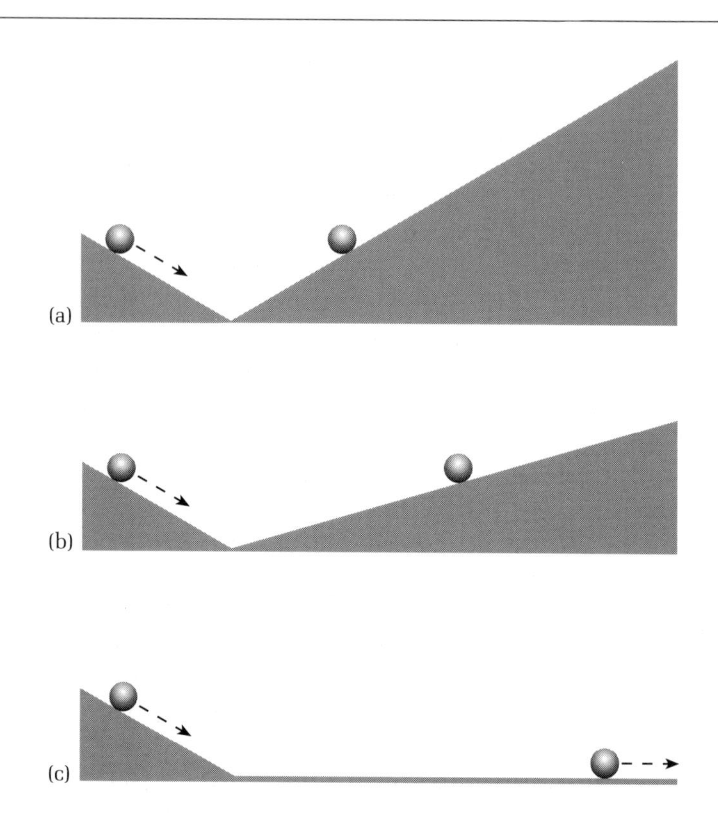


Figure 1-9. Galileo arrived at the concept of the first law of motion through experiments such as the one illustrated here. A ball is rolled down in inclined plane, and it rolls up another plane to the same height it started with (a). As Galileo reduced the inclination of the second plane, the ball travelled farther on it (b). If the plane is horizontal (c), the ball would continue to travel forever, in the absence of friction or an intervening force.

act on a body to generate a constant velocity in it, whereas Galileo's result states that *no* force is acting on the body when it is in the state of constant velocity.

Newton's second law states that the acceleration produced in a body is in proportion to the force applied. And here we encounter another notion that was known qualitatively to Galileo: the notion of *inertia*. Qualitatively, it describes the tendency of any piece of matter to resist a change in its state of motion. Quantitatively, we can say that the greater the inertia of the body, the larger the force that is required to produce the same acceleration in it. Greater force is needed to push a car than to push a bicycle because a car has much more inertia than a bicycle. To put the same idea in a different form, for a given force, we can generate a *greater* acceleration in a body of *smaller* inertia. For the same petrol consumption, a light car accelerates more than a heavy limousine.

Newton ascribed a quantitative measure to inertia through the concept of *mass*. Mass is the quantity of matter in a body. The greater the quantity of matter, the greater is the mass and the greater is its inertia. Using Newton's second law, we can compare the masses of two bodies $A$ and $B$ by simply measuring the accelerations produced in them by the same external force. If the same force produces in $A$ an acceleration twice that of $B$, we conclude that $B$ has twice the mass of $A$.

Newton's third law of motion states that action and reaction are equal and opposite. If we press against a wall, we feel the wall pressing against us. The force we exert on the wall (action) evokes an equal and opposite force (reaction) from the wall on us.

In Figure 1-10, we have a monkey trying to climb a rope over a pulley and carrying a weight equal to that of the monkey. The weight is at the same distance from the pulley as the monkey is. Where will the weight be by the time the monkey reaches the top? With Newton's third law we deduce that the weight will also reach the top at the same time that the monkey does!

## SOME CONCEPTS IN DYNAMICS

Let us now look at the scenario shown in Figure 1-11. This sequence of events could easily have come from a Laurel and Hardy comic movie. A butterfly hits Laurel, causing him a mild annoyance but no more trouble. However, when Hardy, in hot pursuit of the butterfly, hits Laurel, the effect cannot help being spectacular. What property of motion is crucial in producing effects like these?

For the purpose of this illustration, we may assume that Hardy and the butterfly had the same speed, but their effects on Laurel are not the same. Clearly, the large mass of Hardy has made all the difference. But this is not all! If Hardy were walking slowly, he would not have bumped into Laurel so hard. The total effect is

Figure 1-10. A monkey attempts to climb a rope. Whatever pull he exerts on the rope to draw himself up is communicated to the stone by the law of equality of action and reaction. The stone, therefore, moves up in the same way the monkey does.

Figure 1-11. This comic strip illustrates how the momentum of Hardy is communicated to Laurel during their collision.

due therefore to the mass as well as the velocity. The quantity that combines the two is called the *momentum*. Momentum is simply the product of mass and velocity. It has the same direction as that of the velocity.

If we go back to Newton's second law, we now see that it could also be stated in the following form: The rate of change of momentum is equal to the force applied. In an impact with Laurel, Hardy's momentum has clearly changed, the change being caused by the force of impact. And, as Newton's third law implies, Laurel would feel an equal and opposite force of reaction, which is why he is thrown off his chair.

A corollary of Newton's second law is that if there is no net force on a body (or a collection of bodies), the total momentum is unchanged. In the Laurel and Hardy collision, there is no net force – the equal and opposite forces of impact cancel each other and so their total momentum is unchanged. Before impact, Laurel was at rest, while Hardy was moving. Afterwards, most of the momentum is carried by poor Laurel, and because he is lighter than Hardy, he is thrown off the chair with greater speed. This rule of unchangeability of total momentum is called the law of *conservation of momentum*.

A related concept is that of *angular momentum*, and to understand it let us go back to the example of the car on the race track. Suppose the mass of the car is 1000 kg. Then its momentum is simply 1000 (kg) $\times$ 150 (km/h) = 150 000 kg km/h. Its angular momentum about the centre of the track is given by multiplying this momentum by the radius of the track.

In general, a rotating body possesses angular momentum. In Figure 1-12, we see the Earth rotating about its axis. How do we calculate its angular momentum? To do so, we divide the Earth into tiny bits (much as one would divide a picture into jigsaw bits). In Figure 1-12, we see a typical bit moving in a circular track whose centre lies on the axis of rotation. We multiply the momentum of this bit by the radius of the circle along which it moves, just as we did for the race track. We add the contributions from all such bits to get the total angular momentum of the Earth. Like the law of conservation of momentum, we also have a law of *conservation of angular momentum* of a system, provided there are no net forces on the system with a tendency to affect its angular motion. The constancy of angular momentum plays an important role in the dynamical evolution of many astrophysical systems.

Finally, we consider the important concepts of work and energy.

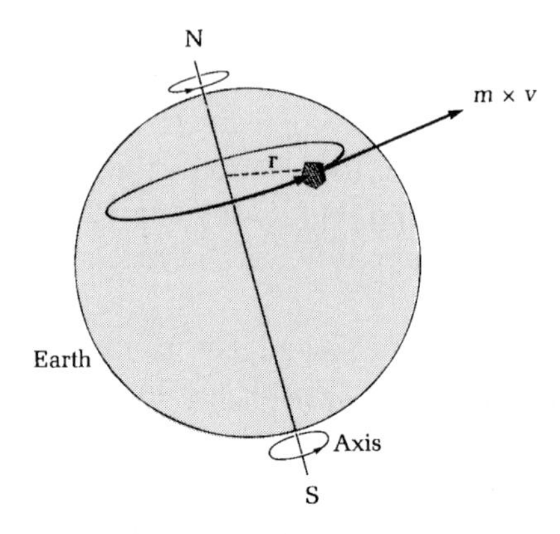


Figure 1-12. A typical bit of matter inside the rotating Earth goes around the axis of rotation (the north–south axis) at a distance $r$. If $v$ is the speed of this bit and $m$ is its mass, it has a momentum $m \times v$ in the tangential direction. The angular momentum of this bit about the axis of rotation is $m \times v \times r$. The angular momentum of the Earth is obtained by adding the angular momentum of all the bits that make up the Earth.

The arguments given by Galileo and Newton have already clarified the relationship between force and motion. Contrary to what Aristotle said, we now see that forces act to change the state of motion rather than to maintain it at a constant velocity. Is there any method of bookkeeping that can tell us what has been achieved by the applied forces at any stage of the motion?

Physicists have found a way of mathematically defining the *work* done by such forces. In Figure 1-13, we see that, by the application of a constant force $F$, a body has been displaced from position $A$ to position $B$. Let $d$ denote the net displacement of the body *along the direction of the force* $F$. Then the product $F \times d$ denotes the *work done by the force.*

Is there any manifestation of this work? We know that the body has accelerated as a result of the application of a force. Suppose it were at rest at $A$ and has attained the velocity $v$ at $B$. A simple calculation using Newton's second law of motion will show that the work done by the force $F$, defined earlier as $F \times d$, is exactly equal to $\frac{1}{2}m \times v \times v$, or $\frac{1}{2}mv^2$, where $m$ is the mass of the body.

This quantity, $\frac{1}{2}mv^2$, is the *kinetic energy* of the body – the energy it acquired by virtue of its motion. So we see that the work done by

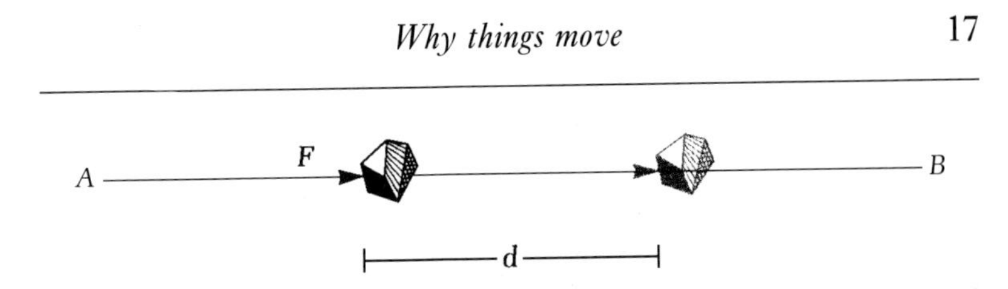


Figure 1-13. The work done by the force $F$ in moving the object over a distance $d$ from $A$ to $B$ is $F \times d$. Only the displacement along the direction of the force counts as work.

the forces has not gone in vain but has resulted in giving an equal amount of kinetic energy to the body.

As we saw in the case of the pushcart brought to a halt by frictional forces, an external force opposing the motion of a body slows the body down. Here we have a case of *reduction* in the kinetic energy of the body. The expenditure of energy has gone toward work done *against* the opposing external force. Thus we can state a general rule that applies to moving bodies

$$\text{change in the kinetic energy} = \text{work done by an external force}$$

If the change is negative (corresponding to a reduction of the kinetic energy), the work is also negative, that is, against the external force. If the change is positive, then the external forces have worked positively in support of the motion. In Chapter 3 we will relate this rule to the law of *conservation of energy*.

## WHY THINGS MOVE

This brings us to the end of our discussion of dynamics, the science of motion, perhaps the oldest of all sciences. Why do things move? An answer to this question was first attempted by Aristotle and was later given in the correct form by Galileo and Newton. Galileo was the first to realize that the effect of forces lies in *changing* the state of motion, while Newton gave quantitative expression to the relationship between force and acceleration.

In nature we see so many different forces. Yet with painstaking advances in theory and experiments, scientists have reduced the basic categories of these forces to only four. In this book we will encounter examples of all four types of basic force, although our main emphasis will be on gravity, the strangest force of them all.

# 2

# *From the falling apple to Apollo 11*

## WHY DID THE APPLE FALL?

Apples have played a prominent role in many legends, myths, and fairytales. It was the forbidden apple that became the source of temptation to Eve and ultimately brought God's displeasure upon Adam. It was the apple of discord that led to the launching of a thousand ships and the long Trojan War. It was a poisoned apple that nearly killed Snow White, and so on.

For physicists, however, the most important apple legend concerns the apple that fell in an orchard in Woolsthorpe in Lincolnshire, England, in the year 1666. This particular apple was seen by Isaac Newton, who 'fell into a profound meditation upon the cause which draws all bodies in a line which, if prolonged, would pass very nearly through the centre of the earth.'

The quotation is from Voltaire's *Philosophie de Newton*, published in 1738, which contains the oldest known account of the apple story. This story does not appear in Newton's early biographies, nor is it mentioned in his own account of how he thought of universal gravitation. Most probably it is a legend.

It is interesting to consider how rare it is to see an apple *actually fall* from a tree. An apple may spend a few weeks of its life on the tree, and after its fall it may lie on the ground for a few days. But how long does it take to fall from the tree to the ground? For a drop of, say, 3 metres, the answer is about three-quarters of a second. So to see an apple fall, we have to be on hand during the crucial short interval of its life! The chance of witnessing this event of course increases if, like Isaac Newton, we sit in an orchard of trees laden with near-ripe apples, but still, the event as such cannot be considered very frequent.

Much less frequent is the appearance of a genius like Newton, who could meditate on such an event and come up with the law of gravitation. Legend has it that Newton's meditations on the question 'Why did the apple fall?' led him eventually to the inverse-square law of gravitation. Newton's answer to this question – 'because the Earth attracted it' – is more profound than it appears to be at first sight, for it not only resolved the mystery of the falling apple but helped resolve a number of long-standing questions about our solar system.

## WHAT IS THE LAW OF GRAVITATION?

Stated in simple words, the law of gravitation tells us that the force of attraction between any two material bodies increases in direct proportion to their masses and decreases in inverse proportion to the square of their distance apart.

Suppose we have two bodies of respective masses $m$ and $M$, situated at a distance $d$ apart. Then the law of gravitation tells us that to find the force $F$ of gravitational attraction between the two bodies, multiply $m$ by $M$ and divide the answer by the square of $d$, that is, by $d \times d$, and multiply the answer by a universal constant $G$. We have already encountered the term *mass*; it is defined as the quantity of matter in a body, and it is also a measure of a body's inertia. We now find another meaning ascribed to it: mass is a measure of how strongly a body can exert a gravitational force on other bodies and also a measure of how susceptible a body is to the gravitational influence of other bodies. In Newton's formula, if we increase $m$ by a factor of 10, the force $F$ is correspondingly increased by a factor of 10. If we decrease $m$ by a factor of 10, the force $F$ is correspondingly decreased by a factor of 10. Because of this property, gravity appears to play a negligible role in the behaviour of atoms and molecules, which have very small masses, whereas it becomes an important force in astronomy, a subject dealing with heavenly bodies of very large masses.

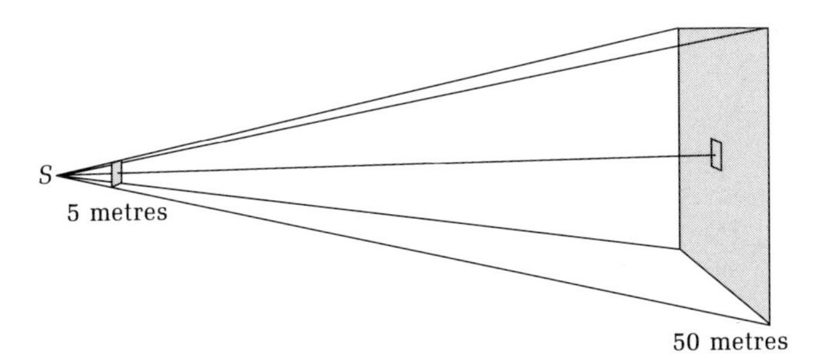


Figure 2-1. A source of light $S$ viewed from 10 times farther away appears 100 times fainter. The amount of light falling on the small square 5 metres away is equal to the amount of light falling on the large square 50 metres away. The area of the large square is 100 times the area of the small square. A small square of the same size at 10 times the distance, therefore, receives only one-hundredth of the light that the nearby square receives.

Because of gravity's diminishing influence with distance, this law is often referred to as the *inverse-square law*. This inverse-square relationship is common in nature. For example, it also applies to the amount of light we receive from a luminous body. If we look at a 100-watt light from a distance of 5 metres, we find it to be very bright. The same light viewed from 50 metres appears faint. Consider a fixed area, as in Figure 2-1, held perpendicular to the path of the light rays. When we increase our distance from the light source by a factor of 10 (from 5 metres to 50 metres), the amount of light we collect on this area per second is reduced by a factor of 100 ($10 \times 10$). The same relationship occurs in the case of $F$, the force of gravity. If we increase the distance $d$ by a factor of 10, the force $F$ is diminished by a factor of $10^2$, or 100.

At this point, it is worth asking, 'Why should gravity be important in astronomy and negligible in atomic physics, when the distances are large in the former and small in the latter?' The answer is that, although according to the inverse-square law the force of gravity should be strong to atomic distances, it is overshadowed by other forces of nature that are considerably more powerful. For example, the force of *electrical* attraction between the electron and proton in a hydrogen atom (Figure 2-2) is estimated to be some-

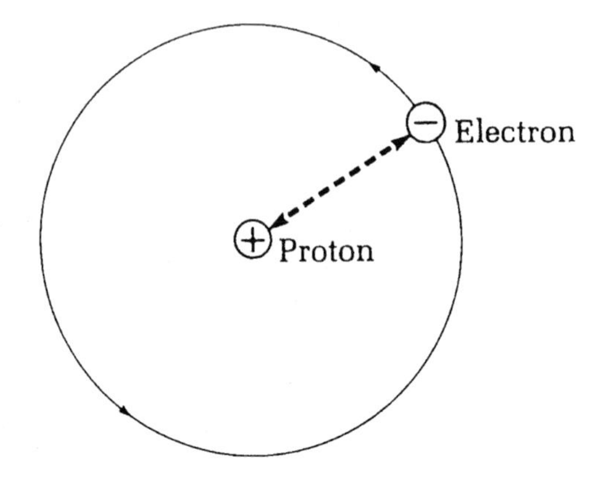


Figure 2-2. The electrical attraction between an electron and a proton in the hydrogen atom is about 10 000 000 000 000 000 000 000 000 000 000 000 000 000 times their gravitational attraction toward each other.

thing like ten thousand billion billion billion billion ($10^{40}$) times the force of their gravitational attraction! Inside the nucleus of the atom, the nuclear forces are even stronger than the electrical forces. The atomic physicist, therefore, rightly ignores gravity in his calculations.

In astronomy, however, none of the other competing forces of nature are able to challenge gravity. The nuclear force is of a very short range; its effect dies out beyond distances of a thousand-billionth of a centimetre! The electrical force does have a long range but since the heavenly bodies are electrically neutral, their electrical force of attraction is zero. This is why astronomers find gravity to be the most important force in their calculations.

Newton argued, on the basis of the law of gravitation, that the apple fell because it was *attracted* by the Earth's gravitational pull. But how did he arrive at the form of the inverse-square law of attraction? Surely, if the purpose of the law was to explain only the falling apple, then it could be served by any law of attraction! In fact, what led Newton to the inverse-square law was not the need to explain the falling apple but the need to explain a much bigger phenomenon – the motion of planets and satellites of our solar system.

## THE MOTION OF PLANETS

In Chapter 1, we saw how Aristotle's ideas had dominated scientific thinking right up to medieval times. Aristotle's ideas led to the so-called *geocentric* theory of the Universe, which assumed that the heavens revolved around the fixed Earth, thus explaining why the Sun and the stars systematically rise in the east and set in the west. We have already mentioned how until the seventeenth century Aristotle's ideas came to be regarded as absolute truths not only in Greece but throughout Europe. The following example from India illustrates how the influence of the geocentric theory had spread well beyond Europe.

The notion of a fixed Earth and the moving cosmos was challenged in India by Aryabhata, a distinguished astronomer of the fifth century AD. In his Sanskrit text on astronomy, the *Aryabhateeya*, there is an explicit mention of the Earth's rotation about its axis (chapter 4, verse 9): 'Just as a man rowing a boat sees the trees on the bank of the river go in the opposite direction, so do the fixed stars appear to us to move from east to west.' However, so well rooted was the notion of 'fixed Earth – moving cosmos' in Indian astronomy that Aryabhata's pupils and successors either denied that he ever held such contrary views or tried to interpret the above verse differently to make it appear less offensive to contemporary scholars. Aryabhata himself had to abandon his native place in north-east India and migrate to the south to evade public ridicule.

The geocentric theory, however, did not stop at describing the motions of stars. The star motions were quite regular and fell within Aristotle's criterion that natural motions are circular or in straight lines. There was another class of objects, the planets, whose motions were considerably irregular. (The Greek word *planaomai* means *wander*.) Some planets, like Venus, showed *retrograde* motion (see Figure 2-3), whereas others appeared to decrease or increase in speed at times. To accommodate such haphazard motions into the Aristotelian system, the Greek astronomers, especially Hipparchus and Ptolemy, made elaborate geometrical constructions involving circular paths known as *epicycles*. In this effort, they were successful to the extent that their *epicyclic theory* could predict with tolerable

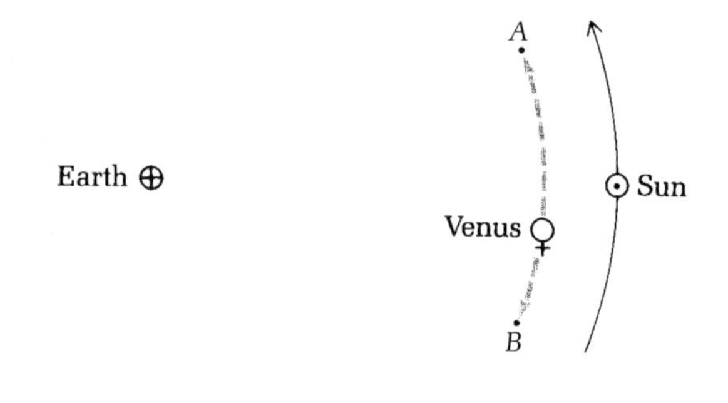


Figure 2-3. As seen from the Earth, the Sun goes around in the direction of the long arrow. Venus also appears to go around, but it sometimes goes ahead of the Sun (to $A$) and sometimes falls back (to $B$). This latter movement is known as retrograde motion.

accuracy in what part of the sky a planet would be found at a given date. The demands on observational accuracy in those days were not so rigorous as they are today, and the successes of this theory naturally raised it to the status of a dogma.

The geocentric theory was challenged by Nicolaus Copernicus (1473–1543), who proposed a rival framework in which to describe the motions in the solar system. This framework, known as the *heliocentric theory*, assumes the Sun to be fixed in space, and the planets, including the Earth, to be orbiting around it. Like Ptolemy, Copernicus also gave elaborate constructions involving circles (a lingering influence of Aristotle?) to describe planetary motions (see Figure 2-4).

The Copernican constructions are simpler but no more accurate than those given by the old geocentric theory. However, their main merit lies in the fact that these constructions, for the first time, pinpoint the central place of the Sun in the planetary system. For someone looking for a dynamical theory – for an explanation of why planets move – the importance given to the Earth in the geocentric theory would be misleading. The clue to the motion of planets lies, as we shall see later, not in the Earth but in the Sun.

The Copernican hypothesis received considerable opposition during Copernicus's lifetime. Copernicus did not see the published version of his book *De Revolutionibus Orbium Caelestium* until he

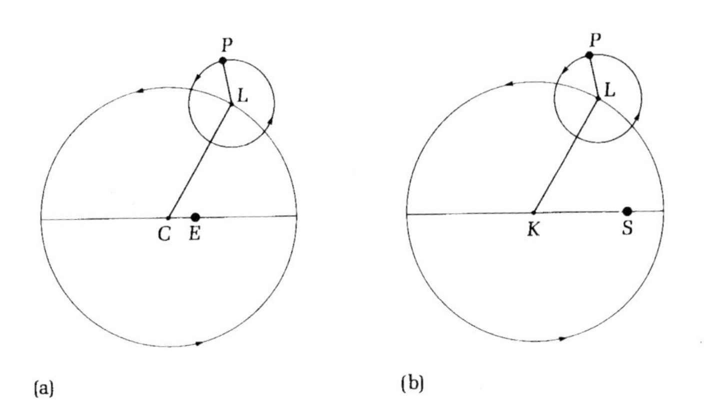


Figure 2-4. The construction of Ptolemy is shown in (a). The Earth is at $E$ and the planet at $P$. $P$ moves on a circle with its centre at $L$. The centre $L$ in turn moves on another circle around $E$ but not centred at $E$. Ptolemy gave elaborate prescriptions for the dimensions of these circles as well as the rates at which the points $P$ and $L$ move on their respective circles. The Copernican construction shown in (b) also involves circles, but now the Sun is identified as the fixed point $S$. The planet $P$ moves on a circle whose centre $L$ moves on another circle *not* centred on $S$.

was on his deathbed. However, its impact on succeeding generations, though gradual, was far reaching.

We have already seen in Chapter 1 how in the seventeenth century the Copernican theory received strong support from Galileo. It was Johannes Kepler (1571–1630), however, whose painstaking observational work marked the next improvement over the Copernican theory. Copernicus had attempted to use circles to describe orbits of planets, but Kepler discovered that these orbits are best described by oblong curves known as *ellipses*. Kepler arrived at the following three laws of planetary motion (see Figure 2-5 for illustration of these laws):

1. The orbit of a planet is an ellipse with the Sun as one of its two foci.
2. The radial line from the Sun to the planet sweeps out equal areas in equal intervals of time.
3. The square of the time taken by the planet to complete one orbit varies in proportion to the cube of the major axis of the orbit.

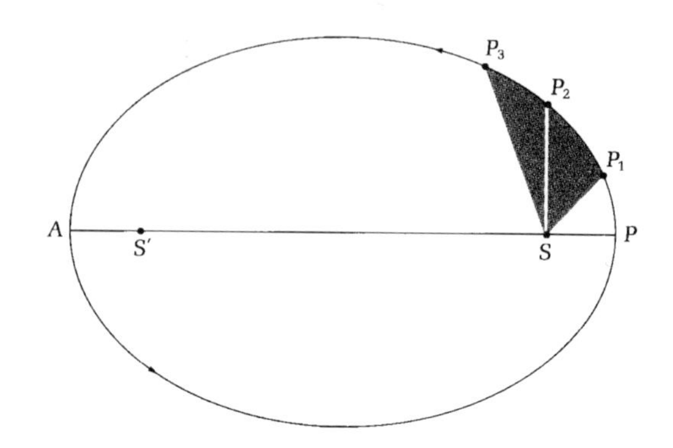


Figure 2-5. The orbit of a planet is best described by an elliptical curve. The Sun $S$ is at one of its two foci, the other focus being at $S'$. The planet moves along the ellipse in such a way that the line joining it to the Sun sweeps out equal areas in equal times. Thus, if the planet goes from $P_1$ to $P_2$ and from $P_2$ to $P_3$ in equal intervals, then the areas $SP_1P_2$ and $SP_2P_3$ must be equal. The line $AP$ is called the *major axis* of the ellipse, with $A$ (the farthest point from the Sun) known as the *aphelion* and $P$ (the point nearest the Sun) known as the *perihelion*. The ratio of $SS'$ to $AP$ measures the *eccentricity* of the ellipse.

Kepler's laws provided the empirical background to Newton's dynamical theory. Kepler's laws described *how* planets move; Newton's laws of motion and gravitation supplied the reason *why* the planets move according to Kepler's laws.

## NEWTONIAN GRAVITY AND MOTION IN THE SOLAR SYSTEM

To draw a circle with a radius $r$ centred at $S$, we attach one end of a string to $S$ and the other end to a drawing pencil $P$. The length of the string between the two ends is $r$. Keeping the string taut we move the pencil around, and it draws a circle. How do we draw an ellipse with foci at $S, S'$ and semimajor axis $a$? The construction is a little more elaborate (see Figure 2-6). Take a piece of string of length $2a$ and attach its ends to $S$ and $S'$. Move a pencil with its end $P$ sliding across the string such that the bits $PS$ and $PS'$ are always taut. In the case of a circle, the pencil end always maintains the distance $PS = a$; for the ellipse, we have $PS + PS' = 2a$.

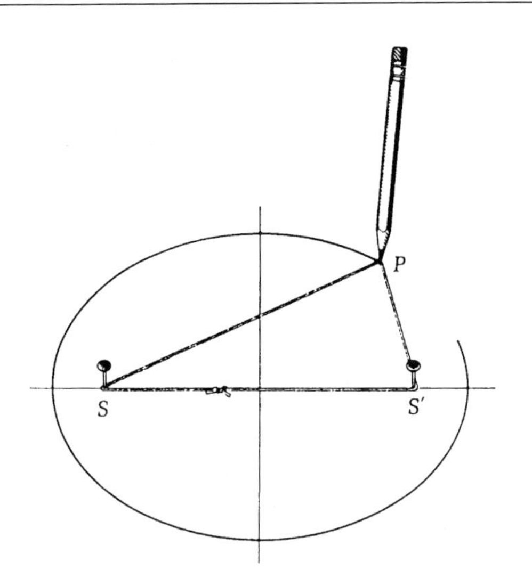


Figure 2-6. How to draw an ellipse. The ratio of the distance $SS'$ to the length of the string $PS + PS'$ is called the *eccentricity* of the ellipse. When the eccentricity is zero, $S$ and $S'$ coincide, and the ellipse becomes a circle.

For the construction of the ellipse, the distance $SS'$ clearly cannot exceed $2a$. If $S$ and $S'$ coincide, the ellipse becomes a circle.

Newton used his system of dynamics to describe the motion of planets pulled by the Sun's gravity. His equations of motion (see Chapter 1) relate the acceleration of the planet to the impressed force, in this case the force of gravity. Knowing the planet's acceleration, can we calculate its actual path in space? To solve this problem, Newton developed a new branch of mathematics which he called *fluxions* but which is now known as the *calculus*. The methods of calculus enabled him to prove that the planets move along elliptical paths satisfying Kepler's three laws. However, the scientific community always tends to be conservative and views new methods with suspicion. To make his theory more readily acceptable, Newton therefore recast his simple proofs based on the calculus into the more conventional but more cumbersome geometrical forms. Newton's book *Philosophiae Naturalis Principia Mathematica* (*The Mathematical Principles of Natural Philosophy*) published in 1687 contains his momentous work on motion and gravitation.

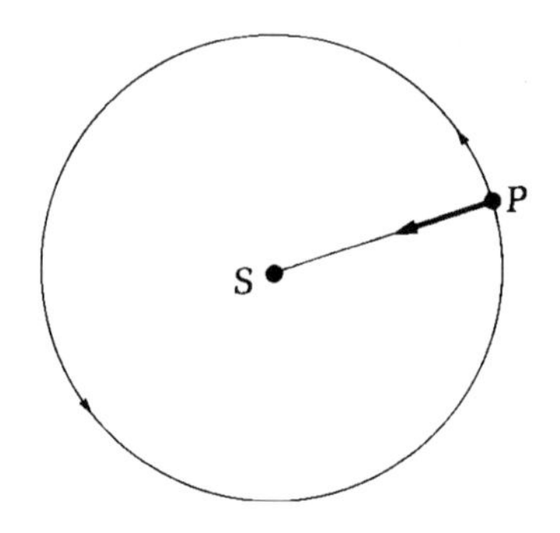


Figure 2-7. The derivation of the inverse-square law of gravitational attraction follows from Kepler's laws of planetary motion. The derivation of the former from the latter is particularly easy in the simplified situation where the planet $P$ moves in a circular orbit with the Sun $S$ at the centre. The force and the acceleration to which $P$ is subjected by $S$ are both along the radial direction shown by the arrow.

Without going into intricate mathematical details, it is possible to see why an inverse-square law of gravitation is implied by Kepler's observations. We will consider the simplified problem of motion in a circle, which we saw above to be a special case of the ellipse.

In Figure 2-7, we have a planet $P$ of mass $m$ moving in a circle about the centre $S$ where the Sun is located. First, we should note that, if the radial line $SP$ is to sweep out equal areas in equal intervals of time (Kepler's second law), $P$ must move with a constant speed along this circle. Suppose the radius of the circle is $r$; then its circumference is $2\pi r$. If the time taken by the planet to go once around this circle is $T$, then its constant speed $v$ must be

$$v = \frac{2\pi r}{T}.$$

In what direction must a force act on $P$ for it to move in a circle? Our first impulse is to argue that the force must be *in the direction of motion*; but this is making the same mistake that Aristotle and his followers made. Force is related *not* to the velocity but to the acceleration. And the acceleration of $P$ is toward the centre $S$ and is equal to $v^2/r$ (see Chapter 1). Thus the force $F$ on the planet must be toward the centre $S$ and is given by Newton's second law, force = mass × acceleration, or

$$F = m \times \frac{v^2}{r}.$$

Since we know $v$ as $2\pi r/T$, we get

$$F = m \times \frac{(2\pi r/T)^2}{r} = \frac{4\pi^2 mr}{T^2}.$$

Now we use Kepler's third law, which tells us that $T^2$ increases in proportion to $r^3$, or

$$T^2 = k\, r^3,$$

where $k$ is a fixed number. Substituting for $T^2$ in the expression for the force $F$, we get

$$F = \frac{4\pi^2 mr}{kr^3} = \frac{4\pi^2 m}{k} \times \frac{1}{r^2}.$$

This tells us that the force on the planet $P$ decreases in inverse proportion to the square of its distance from $S$; that is, it varies in accordance with the *inverse-square law*!

The law of gravitation not only describes the motion of planets around the Sun but also the motion of the Moon around the Earth and the motion of other satellites around their respective planets. That the same law describes the falling apple and the Moon's motion may seem surprising at first. That, like the apple, the moon is *continually falling* toward the Earth is seen with the help of Figure 2-8. There the Moon $M$ is shown as moving in a circle around the Earth $E$. Suppose, by magic, we switch off the force of attraction of the Earth. As shown in the figure, the Moon would then start moving along the dotted straight line with a uniform speed – because there is no force acting on it (Newton's first law)! Compare this trajectory with the Moon's actual circular trajectory around the Earth. Left to itself, the Moon's natural tendency is to break away along the dotted trajectory; but the Earth is constantly pulling the Moon toward it. Hence we can look upon the Moon's motion as if it is continually falling toward the Earth. Because it has transverse velocity, it never actually reaches the Earth but keeps moving around it.

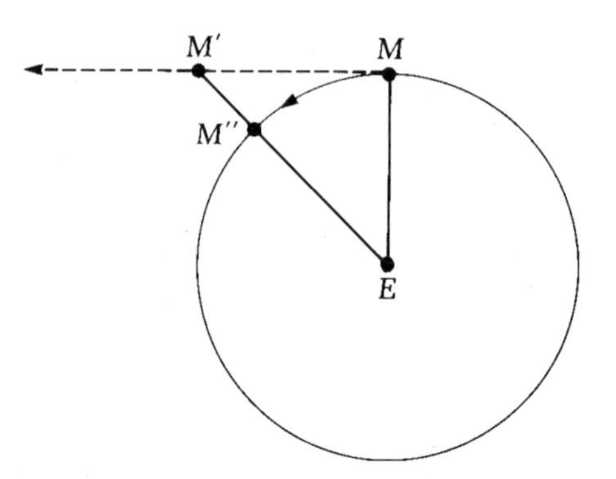


Figure 2-8. In the absence of the gravitational pull of the Earth, the Moon would have moved from $M$ to $M'$ along the dashed straight line. Because of gravity, the Moon actually moves along the circular arc $MM''$. Relative to the Earth $E$, the circular arc is closer than the straight line. Viewed in this light, the Moon, as it moves in the transverse direction, also falls *toward* the Earth (in this case from $M'$ to $M''$).

These ideas, therefore, bring out the *quantitative* role of astronomical observations in the deduction of the law of gravitation. Simply observing falling apples would not have given Newton the clue that the force of attraction follows the inverse-square law. That important input came from Kepler's analysis of planetary motion.

## WHO FIRST THOUGHT OF GRAVITATION?

Newton was not, however, the first to have thought of gravitation. As early as the fifteenth century, some astronomers had the idea that a force of attraction might exist between heavenly bodies and the Earth. It was argued that the Earth is being pulled in all directions by a 'magnetic' force, but since the force is the same in all directions, the Earth remains at rest.

Gilbert in 1600, Ismaelis Bouillard in his book *Astronomica Philolaica* published in 1645, and Alfonso Borelli in 1666 appear to have come close to the basic features of the Newtonian law of gravitation, as did Kepler, who once actually considered the inverse-square law before rejecting it.

The apple legend credits Newton with the idea of gravitation in

1666, although his first publication on it, a treatise called *Propositions de Motu*, was communicated to the Royal Society in February 1685, while the *Principia* itself was published in 1687. In the meantime, in 1674, Robert Hooke published his work describing the motion of the Earth around the Sun in terms of a law of attraction that decreased with distance. Hooke is said to have communicated his ideas to Newton, who had also arrived at similar conclusions independently.

Why did Newton wait for so long – nearly two decades – before publishing his results? In the present scientific era of 'publish or perish', where rushing to the media for announcement of half-baked results is not uncommon, it becomes all the more difficult to understand Newton's reticence.

It is argued that Newton was a perfectionist and wanted to wait until he had sorted out some problems connected with his theory. One of these problems was the need for a mathematical proof that a spherical body attracts others as if its mass were concentrated at its centre. This problem is illustrated in Figure 2-9. The other problem was an observational one. It seems that Newton wanted to wait until reliable estimates of the dimensions of the Earth–Sun–Moon system became available so that he could test the correctness of his theory. These became available in the late 1670s. It was only then that Newton felt confident in his law of gravitation.

Controversy still exists about why Newton waited and about the extent of credit that Hooke should be given for the law of gravitation. About the finished product, however, there is no doubt. The credit for the mathematical computations of planetary orbits based on the laws of motion and gravitation goes to Newton. None of his contemporaries had the mathematical expertise or the breadth of knowledge to carry through such a calculation.

Newton's dislike for controversy and his reticence are reflected in his communication to Edmund Halley when submitting Book II of the *Principia* for publication. By then, Book I of the *Principia* had been published, and Hooke had begun a dispute, claiming priority in the authorship of the law of gravitation. Halley was acting as the peacemaker in the controversy. Referring to Book III (to follow Book II), Newton wrote: 'The third I now design to suppress. Philosophy is such an impertinently litigious lady that a man had as

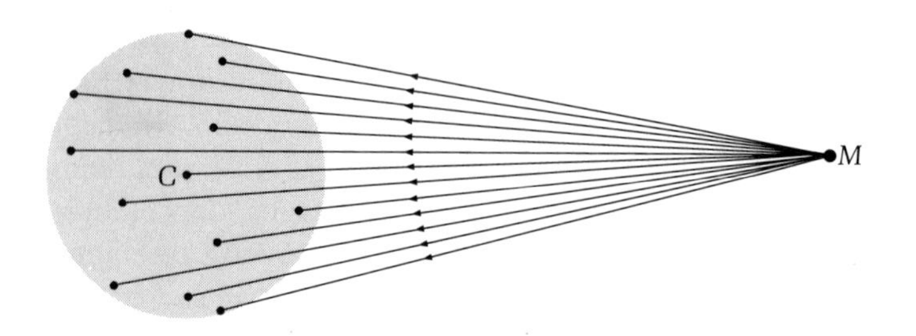


Figure 2-9. If we divide a sphere into tiny bits, each bit will gravitationally attract a particle $M$ toward itself. A few such bits are shown in the figure. The result of *all* these forces on $M$ will be a force acting in the direction $MC$, where $C$ is the centre of the sphere. The magnitude of the force will be what it would be if the entire mass of the sphere were concentrated at $C$. This was the result which Newton wanted to prove mathematically before applying his law of gravitation to spherical planets and satellites as well as the Sun.

good be engaged in law-suits to have do with her.' However, to the advantage of posterity, Halley succeeded in persuading Newton to change his mind. The third volume of the *Principia* duly appeared.

## SUCCESSES OF THE LAW OF GRAVITATION

Leaving aside the controversy about who should be credited with the genesis of the law of gravitation, let us now review some of its achievements. The law of gravitation implied instantaneous action at a distance. The force of gravity between the Sun and the Earth is communicated instantly across a distance of some 150 million kilometres. How is this done? Why did the attraction diminish according to the inverse-square law? Questions like these troubled Newton's contemporaries and successors. When asked about such questions, Newton is believed to have said 'Non fingo hypotheses' (I do not feign hypotheses). Newton attached more importance to the requirement that the law should adequately describe observations than to speculations about nature's mysterious processes leading to that law.

Indeed, it was the successes achieved by Newton's law that established it so firmly in post-Newtonian physics. The bothersome deeper questions of how and why were relegated to the background

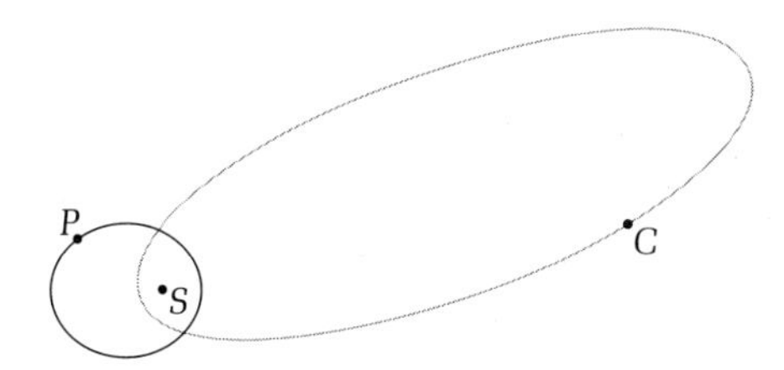


Figure 2-10. Typical orbits of a planet $P$ and a comet $C$. Both orbits are ellipses with the Sun $S$ at the focus. The planetary orbit is nearly circular, while the orbit of a comet is a highly eccentric ellipse.

by the successes of the inverse-square law. Let us look at some of its triumphs.

The first example is that of Halley's comet. Like a planet, a comet also moves in an elliptical orbit because of the Sun's gravitational pull. However, unlike a planet, a comet moves in a highly eccentric orbit. If we go back to our construction of an ellipse, we recall that a highly eccentric ellipse will result if the separation between $S$ and $S'$ is very nearly equal to (but still less than) the length $2a$. An example of a cometary orbit as distinct from that of a planet is shown in Figure 2-10. Notice that the length of the major axis of a comet's orbit is very long compared to that of the orbit of a planet.

As a result of moving in such an orbit, the comet is periodically seen in the vicinity of the Sun after long intervals. But, unless the orbit of the comet (which extends to remote parts of the solar system) is disturbed by an intervening planet like Jupiter, its visits to the neighbourhood of the Sun are with a fixed period.

Edmund Halley, a contemporary and friend of Newton, noticed that comets had been sighted at regular intervals, and argued that these were not different comets but periodic visitations of the *same* comet. Thus, for a comet that was seen in 1682 Halley argued that this was the same comet that had come earlier in 1456, 1531, and 1607 – at a regular interval of around 76 years. Halley predicted that it would be seen again in 1758. This prophecy came true, although Halley did not live to see the comet's passage in that year.

Halley's comet came in 1986, again as expected. This time, for the first time in its history, technology had advanced to such a level that several nations were able to launch spacecraft to rendezvous with the comet. These spacecraft came near the comet and took pictures from close by (see Figure 2-11). Even the launching and routing of the spacecraft in such orbits could be done as precisely as they were because of the correctness of Newtonian laws as applied to these craft.

Perhaps no one did more to establish confidence in the law of gravitation than Pierre Simon Laplace (1749–1827), the French mathematician. Laplace's five-volume work *Mecanique Celeste*, published from 1799 to 1825, has been compared with Ptolemy's *Almagest* for its impact on contemporary astronomy. In this work, Laplace applied the latest mathematical techniques to work out the motions of planets and their satellites under each other's gravitational influence. The problem is extremely intricate when one takes into account all the cross-influences of the eighteen bodies (then known) of the solar system. Faced with such a problem in modern days, the inclination of the physicist would be to 'put it all on a computer'. The success achieved by Laplace in solving his mammoth problem, and the resulting agreement between his calculations and the observations of planets and satellites, convinced the skeptics about the validity of Newton's law of gravitation. When Laplace presented his work to Emperor Napoleon, he looked at the contents and asked Laplace why his book made no mention of God. Laplace is said to have replied, 'Sire, I had no need of that hypothesis.'

The next triumph of Newtonian theory came in 1845, when it was used in the discovery of a new planet. Two astronomers, Adams in England and Leverrier in France, came to this discovery working independently. Their work was based on the irregularity that had been noticed in the orbit of Uranus, then the farthest-known planet of the solar system. Uranus was apparently not following the exact elliptical orbit predicted by Newtonian gravity. Both Adams and Leverrier concluded that the irregularity in the motion of Uranus was caused by a new planet in its vicinity; the gravitational pull of this new planet would be responsible for the perturbation of Uranus's orbit. The two astronomers were able to calculate where the new planet should be located. Adams approached Challis, the Director of the Cambridge Observatory, and Airy, the Astronomer Royal, with the request to observe this planet. But his requests

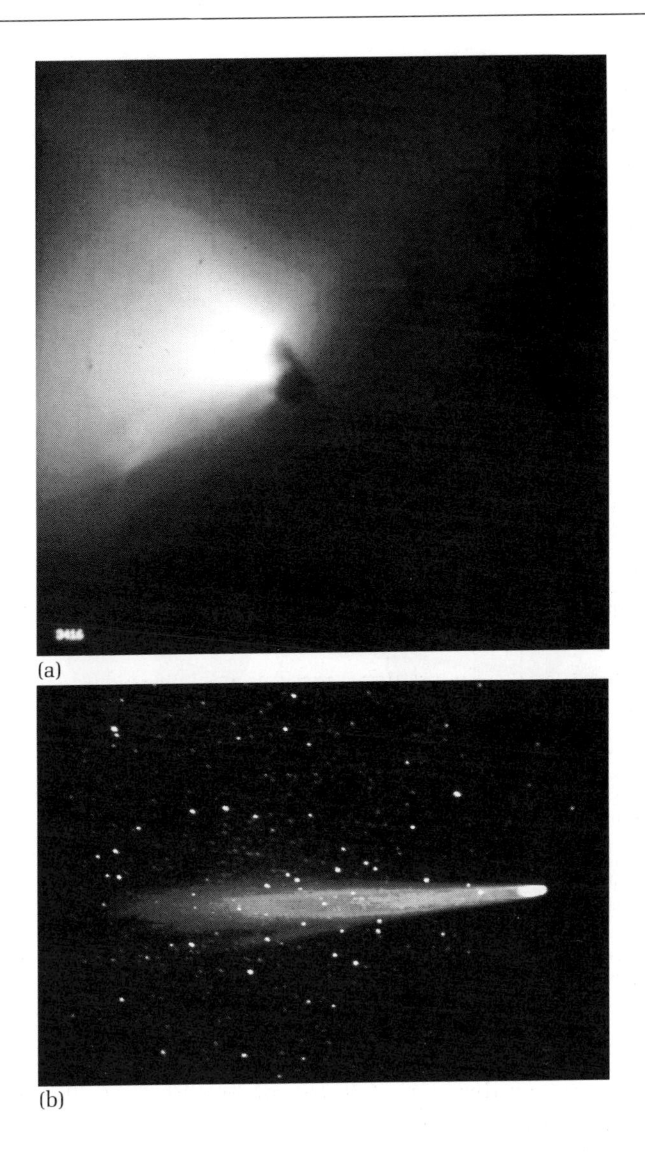

Figure 2-11. (a) The photograph of Halley's comet in 1986 taken from the Giotto spacecraft that was launched by the European Space Agency to rendezvous with the comet. (By courtesy of H. U. Keller, copyright Max-Planck Institut für Aeronomie, Lindau, Germany.) This is placed above (b) the photograph of the comet taken in 1910. (Courtesy of Mount Wilson and Las Campanas Observatories, Carnegie Institution of Washington.)

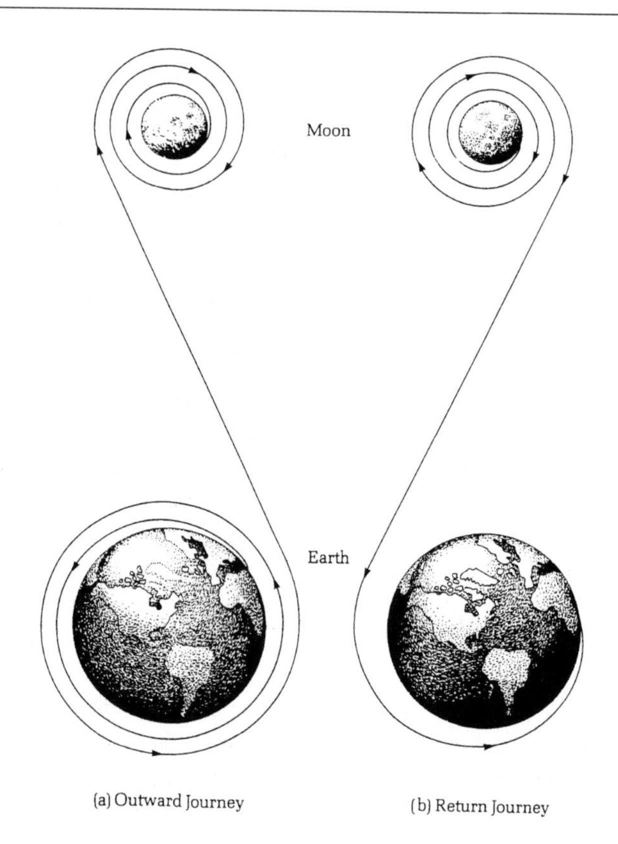


Figure 2-12. A schematic diagram of the orbit of a spaceship from the Earth to the Moon and back. The dynamics of this complicated series of motions is governed by Newton's laws.

were ignored. Leverrier fared no better with the leading French astronomers. However, his work was taken seriously by Galle of the Berlin Observatory, who succeeded in observing and locating the new planet Neptune. The story has it that the Director of the Berlin Observatory was on leave celebrating his birthday, and Galle, a younger astronomer, happened to be around when Leverrier's request came. Had the Director been present he perhaps might not have been sympathetic to it. The Neptune episode illustrates the fact that, if a scientific theory follows the right lines, apparent disagreement with its predictions can lead to new scientific discoveries. It also tells us that senior scientists should take their junior colleagues seriously!

While these three examples relate to the natural constituents of the solar system, our fourth and last example deals with artificial satellites and spaceships. These objects – whether the first Sputnik to go around the Earth, or the Apollo 11 mission to the Moon, or the Viking, Pioneer, and other space missions to other planets or, as we saw in the case of Comet Halley, a space mission to a comet – their motion is governed by the same law of gravitation that Newton gave three centuries ago (see Figure 2-12).

For example, the Apollo 11 journey from the Earth to the Moon (and back!) had to take into account the following motions. First, there is the Earth's motion around the Sun and the Moon's motion around the Earth. In fact, we have a 'three-body problem' in which each body moves under the gravitational pulls of the other two. Next, the motion of the spaceship from the Earth to the Moon is governed by the gravitational pull of the Earth and the Moon on the spaceship. This calculation of the correct trajectory is complicated and can be done with ease only on an electronic computer.

The accuracy of present-day space missions is considered to be a triumph of modern technology. It is also a vindication of the law of gravitation, allegedly inspired by a falling apple. It is therefore with some confidence that we next consider even more remarkable manifestations of gravity in astronomy.

# 3

# *How strong is gravity?*

## THE MASS OF THE EARTH

Although with Newton's pioneering discoveries, gravity was the first basic force of nature to be described and studied quantitatively, it is the weakest of all known basic forces of nature. The other basic forces are the forces of electricity and magnetism and the forces of 'strong' and 'weak' interaction which act on subatomic particles. It is a measure of the success achieved to date that physicists are able to explain all observed natural and laboratory phenomena in terms of these four basic forces. As we shall see in later chapters, many physicists hope that one day they will be able to bring all the basic forces under the umbrella of one unified force.

Although atomic physicists consider gravity to be the weakest of the four known basic forces of nature, for astronomers gravity is the most dominant force in the celestial environment. How do we assess the *strength* of gravity in any given situation? We will try to answer this question with a few examples in this chapter.

All of us on the Earth are conscious of gravity. The feeling of weight that we have results from the gravitational pull the Earth exerts on us. Newton's inverse-square law of gravitation described in Chapter 2 tells us how strong this force is on any given body on the Earth's surface. Let $m$ be the mass of the body and $M$ the mass of the Earth. The distance between the body and the Earth is denoted by $d$. Newton's law then tells us that the force of attraction between the body and the Earth is given by

$$F = G \times \frac{m \times M}{d^2}.$$

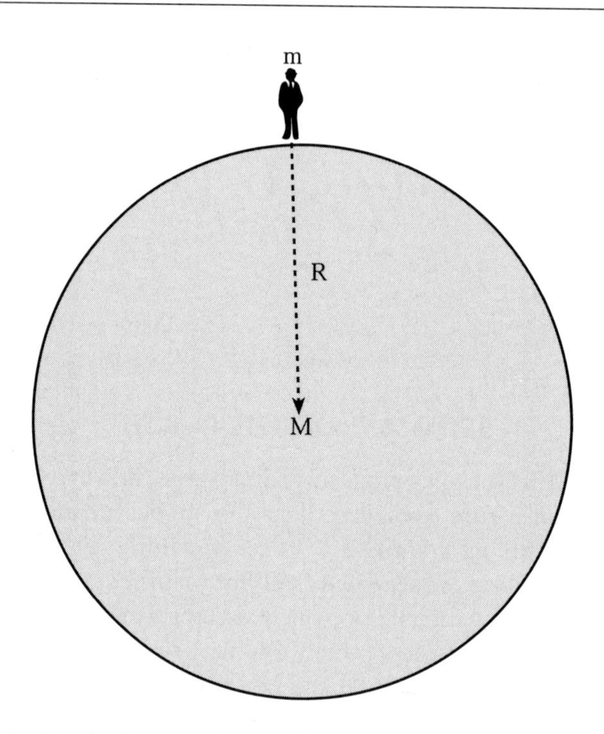


Figure 3-1. A body of mass $m$ close to the surface of the Earth is attracted as if the entire attracting mass $M$ of the Earth (assumed to be spherical here) is concentrated at the centre of the Earth.

Let us examine what we know about $d$ and the gravitational constant $G$.

What is $d$? As shown in Figure 3-1, the Earth is nearly a perfect sphere with a radius of about 6400 kilometres. From where on Earth should we measure the distance to the body? The distance is zero from the point on the surface where the body lies, while it is 12 800 kilometres from the diametrically opposite point. The distance of any other point in the Earth will lie somewhere between these extreme values.

Here we recall a result from Chapter 2, a result that is said to have taken Newton several years to prove: *A spherical body attracts as if its entire mass were concentrated at its centre.* For a spherical Earth, the correct value for $d$ is therefore equal to $R$, the Earth's radius – about 6400 kilometres.

We now come to an interesting consequence of the above result. If we know the value of $G$, the constant of gravitation, we can measure *the mass of the Earth*! Henry Cavendish (1731–1810) was the first to give an experimentally measured value of $G$. Of course, modern techniques give us the value of $G$ much more accurately than the early measurements of Cavendish.

Let us see how we can measure $M$, the mass of the Earth, with this information. First, recall the experiment performed by Galileo at the Leaning Tower of Pisa (Figure 3-2). Galileo demonstrated that all bodies dropped from a height fall with equal velocity. This experiment is easily interpreted in terms of Newton's second law of motion, force = mass × acceleration. We have already seen what the force of gravity is on a body on or close to the surface of the Earth. Dividing that force by the mass of the body gives us the acceleration. Denoting acceleration by the symbol $g$, we get the simple formula

$$g = G \times \frac{M}{R^2}.$$

Notice that $g$, the acceleration of the falling body, *does not* depend on its mass, which explains Galileo's conclusion, that all bodies fall with equal velocity.* The value of $g$ is approximately 9.81 metres per second per second.

We can write this relation in a slightly different form:

$$M = \frac{g \times R^2}{G}.$$

When it is written this way, we have *all known quantities* on the right side of this relation: $g$, the acceleration of the falling body; $R$, the radius of the Earth; and $G$, the constant of gravitation. A simple calculation then gives us the *mass of the Earth* as approximately six thousand million million million tons. The enormous size of this mass by our everyday standards should clarify why the force of gravity is felt by all of us. The Earth is the single most massive object in our environment.

This same method used for measuring the mass of the Earth can

* In arriving at this conclusion, *air resistance* as a force is neglected. A sheet of paper dropped from a second-storey window does *not* fall as rapidly as a pen dropped from the same place because the air resistance is considerably higher on the paper than on the pen. Galileo was aware of this effect.

Figure 3-2. Galileo dropped objects of various shapes and sizes from the top of the Leaning Tower of Pisa to demonstrate that all bodies fall with equal speed. The tower is about 55 metres high and has a maximum inclination from the vertical of about 4 metres, or just over 4°.

also be used for measuring the mass of the Sun. In Chapter 2, we saw that the Moon moving around the Earth is also continually falling toward it. Likewise, the planets are continually falling toward the Sun. It is therefore possible to think of the acceleration of the Earth as it 'falls' toward the Sun. Thus, the formula that enabled us to measure the mass of the Earth can also be used to estimate the mass of the Sun. All we have to do is substitute for $g$ the acceleration of the falling Earth and for $R$ the radius of the Earth's orbit. The mass of the Sun turns out to be about two thousand million million million million tons. Thus the Sun is about 330 000 times as massive as the Earth!

## MASS AND WEIGHT

A common mistake in everyday language sometimes causes confusion between the concepts of *mass* and *weight*. Mass is the quantity of matter contained in a body. As we saw in Chapter 1, mass is the measure of a body's inertia, the property by virtue of which the body resists any change in its existing state of rest or motion. In Chapter 2, we found another property of mass – it measures the strength with which the body attracts and is attracted by other bodies. These properties of mass will continue to apply to the body without any quantitative change *no matter where it is in the Universe*; mass is an *intrinsic* property of the body.*

The *weight* of a body, on the other hand, measures the *force* with which it is gravitationally attracted by the Earth. If the body is taken from the Earth to some other planet, its weight will be given by the force with which that planet attracts it. Weight can therefore vary, depending on the location of the body. Even on the Earth, the weight of the same body can vary from place to place. Because of the rotation and the flattening of the Earth at the poles, a person will weigh 0.25% more at the poles than at the equator.

The weight of a body on the Moon is nearly one-sixth of its weight on the Earth (see Figure 3-3), because the value of $g$, the acceleration due to gravity on the Moon, is one-sixth of that on the Earth. Table 3-1 shows how the weight of a person will vary on different planets.

## GRAVITY BARRIERS

Because of one's reduced weight on the Moon, any person will be able to jump higher and throw a ball farther on the lunar surface than on the Earth. Or, to put it differently, greater effort is needed to throw a ball to the same height on the Earth than on the Moon. Physicists express this fact by saying that *gravity erects a taller potential barrier on the Earth than on the Moon.* Let us try to understand this statement with the help of an example.

* Ernst Mach has questioned the validity of the Newtonian concept that mass is an intrinsic property of a body. In his book *The Science of Mechanics*, published in 1893, Mach gives persuasive arguments against this concept. Mach's ideas have had considerable influence on the thinking of many physicists in this century, including Albert Einstein. We will come back to 'Mach's principle' in Chapter 12.

Figure 3-3. A body has the same mass but different weight from location to location, depending on the force of gravity. One's weight on the Moon is one-sixth of one's weight on the Earth!

Table 3-1.

| Planet | Weight as a percentage of weight on Earth |
|---|---|
| Mercury | 37 |
| Venus | 89 |
| Earth | 100 |
| Mars | 38 |
| Jupiter | 265 |
| Saturn | 114 |

In Figure 3-4, we see a man attempting to toss a ball into a basket located vertically above him. His job is an easy one as long as the basket is located only a few metres above his head. However, as the basket is raised to greater and greater heights, the man has to throw the ball up with greater and greater speed. In part b, we see what happens when the ball is thrown up at a speed of 12 metres per

Figure 3-4. (a) Because Earth's gravity pulls the ball down, the thrower has to hurl it up with a certain minimum speed if the ball is to reach the basket. This minimum speed increases with the height of the basket above the ground. (b) The speed of the ball decreases as it rises. The dashed line on the graph shows this decreasing trend of speed with height.

second. The graph shows how the speed of the ball is progressively *reduced* as it rises. Starting with the value of 12 metres per second from the point of ejection, the speed declines steadily until it falls finally to zero at a height of about 7.5 metres.

This 7.5 metres is the maximum height attained by the ball in the above example. If it did not reach the basket, it would begin to fall with increasing speeds. It would attain the same sequence of speeds at the corresponding heights that it had while rising. This relationship between height and velocity can be understood in terms of the energy–work relation discussed in Chapter 1. When the ball was rising, it was doing work *against* the force of gravity. The work done against gravity resulted in a decrease in the *kinetic energy* of the ball. By the time the ball attained the height of 7.5 metres, it could no longer draw upon its store of kinetic energy. It stopped rising and came to rest.

At this point, we may wonder if there has been a net loss of the ball's energy. Certainly, the ball had been progressively losing kinetic energy as it rose. This energy was lost in doing work against the gravitational pull on the Earth. Did the ball have anything to show for it when it reached the maximum height of 7.5 metres? The answer is yes. The ball begins to acquire *potential energy* as it rises in just the amount that it loses kinetic energy. This potential energy increases in proportion to the height of the ball, being at a maximum when the ball is at 7.5 metres. The word 'potential' indicates that the ball acquires the possibility to make gravity do work by virtue of its height. As the ball descends, its kinetic energy *increases* because gravity does work on it, while its potential energy *decreases* by an equivalent amount. In a hydroelectric dam (see Figure 3-5), potential energy is converted to another form of energy, electricity.

We can therefore arrive at the *law of conservation of energy* for the moving ball:

$$\text{kinetic energy} + \text{potential energy} = \text{constant}.$$

Throughout the ball's motion, whether it is going up or down, the total of the two forms of energy (kinetic and potential) is constant.

Having reassured ourselves that the ball's total energy is conserved, let us return to the ball thrower. His muscular power limits him to a certain speed above which he cannot throw the ball. This circumstance, therefore, puts a limit on the height to which he can throw the ball. We saw previously that the maximum height is about 7.5 metres for a speed of 12 metres per second. A stronger thrower can of course aspire to greater heights.

Figure 3-6 shows how the maximum height varies with the thrower's beginning speed. Although the height attained increases

Figure 3-5. The Bhakra Nangal hydroelectric dam in north India converts the kinetic energy of water, acquired as a result of falling from a height, into electrical energy. With a height of about 200 metres, it is one of the tallest straight gravity dams in the world. (Photograph courtesy of the Bhakra Beas Management Board, reproduced by permission of the Ministry of Energy, Department of Power, Government of India.)

with the speed of the throw, in each case the ball would come back to the ground if there were no basket to hold it up. No matter how strong the ball thrower is, therefore, it appears that he cannot throw the ball high enough to escape the Earth altogether!

This brings us to the notion of a *gravity barrier*. To launch a spaceship to the Moon or to more distant parts of the Universe, we have to make sure that it leaves the confines of the Earth. We may imagine the spaceship to be situated at the bottom of a 'gravity well'. Escape from the Earth is then equivalent to climbing the wall of this well – that is, to doing sufficient work to surmount the barrier erected by the Earth's gravity. Just how tall is the Earth's gravity barrier?

## ESCAPE SPEED

In our previous example of the ball thrower, we have assumed the downward acceleration due to gravity to be $g$ = 9.81 metres per

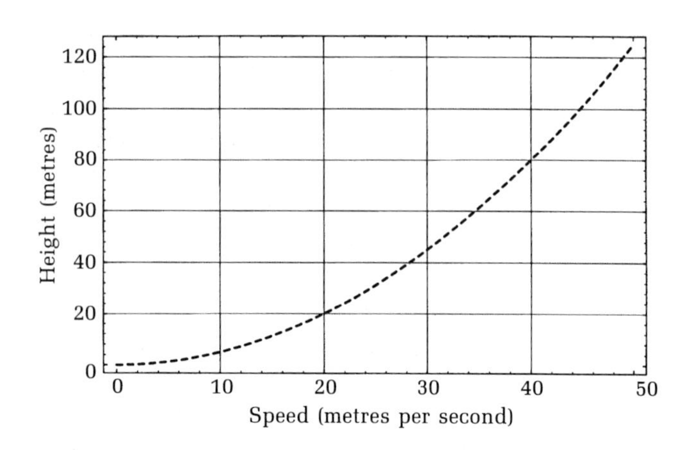


Figure 3-6. The variation of the maximum height attained with the speed of the ball. If the dashed curve were continued well beyond the range of the graph, it would tell us that, to attain a height of 6400 kilometres, the speed needed is about 11.2 kilometres per second. This is the escape speed for a projectile fired from the Earth's surface.

second per second. This is the value of $g$ on the Earth's surface, a value we first saw in connection with Galileo's experiment at the Tower of Pisa. This is, however, one effect which we have hitherto ignored but which becomes important as we progressively increase the height attained by our basketball.

Let us take an example. Suppose the ball thrower has a maximum range of 30 metres above the ground. Although the strength of the Earth's attraction decreases as the ball ascends, the rate of the decrease is so slow that over this height the value of $g$ is more or less the same. So our conclusion about the relation between the maximum height attained by the ball and the initial throwing speed is not wrong; Figure 3-6 correctly describes this relationship.

However, when we extrapolate this relation to space travel, we must remember that, as we go farther and farther from the Earth, *there is a significant decrease in the force of attraction.* Thus the value of $g$ at a height of 6400 kilometres (the radius of the Earth) is *one-quarter* of the value on the surface of the Earth. That is, $g = 2.45$ metres per second per second at this height. Hence the work done in climbing 30 metres at an altitude of 6400 kilometres is one-quarter of the work done in climbing the same height at the surface of the Earth. The depth of the Earth's gravity well is therefore less than what we previously estimated. Mathematical calculation shows that the work required to lift a unit of mass out of this well is not infinite but is equal to $GM/R$, where $M$ and $R$ are, as before, the mass and radius of the Earth and $G$ is the constant of gravitation. When we recall that on the surface of the Earth $g = GM/R^2$, this work per unit mass can be simply expressed as $g \times R$. In other words, the work done to lift a spaceship out of the Earth's gravity well is equal to climbing a height of 6400 kilometres *as if the value of* g *did not change from its surface value.*

Referring back to our previous discussion and to Figure 3-6, we therefore ask, what is the value of the ball thrower's speed of throw for the ball to rise to a height of 6400 kilometres? The answer is, nearly 11.2 kilometers per second, or about 40 000 km/h. This is the minimum speed that will carry a spaceship beyond the confines of the Earth's gravity. It is known as the *escape speed* for the Earth's surface.

## HOW STRONG IS GRAVITY?

The escape speed gives us an indication of the strength of gravity on the surface of the Earth, and we can use this concept to compare the strength of gravity on various astronomical objects. In general, for any spherical object of mass $M$ and radius $R$, the escape speed is given by the same formula that we used for the Earth,

$$V = \sqrt{\frac{2GM}{R}}.$$

On the surface of the Moon, the escape speed is only 2.4 km per second. This means that it is considerably easier for a spaceship to leave the lunar surface than it is for the same spaceship to leave the Earth. Table 3-2 gives the escape speeds for several astronomical bodies.

The stronger the gravitational pull of the object, the larger is the escape speed. In Table 3-2, the neutron star has the largest escape speed, more than half of the speed of light. A neutron star is a very

Table 3-2.

| Astronomical object | Escape speed (kilometres per second) |
|---|---|
| Moon | 2.4 |
| Earth | 11.2 |
| Jupiter | 60.8 |
| Sun | 640 |
| Sirius B (a white dwarf star) | 4 800 |
| Neutron stars | about 160 000 |

compact star; its density is a million billion times that of water. A neutron star as massive as the Sun may have a radius of only 20 kilometres!

Does the neutron star hold the record for the greatest speed necessary to escape from any object? Theoretical physicists tell us that, in principle, the greatest possible speed of a physical body cannot exceed the speed of light. For an object whose radius is determined by the formula

$$R = \frac{2GM}{c^2} \quad (c = \text{ speed of light}),$$

*the escape speed equals the speed of light.* Such an object, if it exists, should hold the above record! In other words, even light can barely escape from the surface of such an object. Astronomers rely on electromagnetic radiation in its various forms – visible light, radio waves, X-rays – to observe objects, but they cannot observe an object if its escape speed is faster than the speed of light. Such an object can only be detected by its strong gravitational force.

This is our first encounter with the concept of a *black hole.*

# 4

# *Fusion reactors in space*

## FUSION

It is often argued that man's growing energy needs will be met if he succeeds in making fusion reactors. In a fusion reactor, energy is generated by fusing together light atomic nuclei and converting them into heavier ones. The primary fuel for such a fusion reactor on the Earth would be heavy hydrogen, whose technical name is deuterium. Through nuclear fusion, two nuclei of deuterium are brought together and converted to the heavier nucleus of helium, and in this process nuclear energy is released.

The following is the recipe for a fusion reactor. First, heat a small quantity of the fusion fuel, deuterium, above its ignition point – to a temperature of some 100 million degrees Celsius. Second, maintain this fuel in a heated condition long enough for fusion to occur. When this happens, the energy that is released exceeds the heat input, and the reactor can start functioning on its own. The third and final part of the operation involves the conversion of the excess energy to a useful form, such as electricity.

The primary fuel for this process, the heavy hydrogen, is chemically similar to but a rarer version of the commonly known hydrogen. An atom of ordinary hydrogen is made up of a charged electrical particle called the *proton* at the nucleus with a negatively charged particle, the *electron*, going round it. The nucleus of heavy hydrogen carries an additional particle called the *neutron* in its nucleus. The neutron has no electric charge so the total charge of the nucleus of heavy hydrogen is the same as that of ordinary hydrogen.

Just as the molecule of ordinary water is made of two atoms of hydrogen and one atom of oxygen, we get a molecule of *heavy water* if we replace the hydrogen atoms by deuterium ones. Sea water contains a small fraction of heavy water. Thus the fuel of our nuclear reactor can be extracted easily and abundantly from the heavy water available in the oceans.

Thus the recipe appears simple, but it is not so simple to translate into reality. World wide research efforts over many years have yet to yield a practical solution. Nuclear fusion as such was achieved years ago when hydrogen bombs were made. The H-bombs are a testimony to the vast reservoir of energy in atomic nuclei that can be released through fusion. Where then is the difficulty in making fusion reactors?

The difficulty lies in achieving *controlled* nuclear fusion. We want the nuclear energy to be released steadily and not explosively as in a bomb. For this to occur, the fusion material must be properly confined and held in a stable form. This is the crux of the problem that present research is trying to solve.

The first indication that nuclear fusion can operate in a controlled fashion and generate useful energy came not from any laboratory demonstration on the Earth but from the study of stellar structure, from human curiosity about the secret of stellar energy. Ever since they gazed at the night sky full of stars, human societies have wondered what makes the stars shine. Shining requires the generation of light, and light is a form of energy. Where do the stars get their energy?

While looking for a clue to the stellar energy source, astronomers hit upon the concept of nuclear fusion. They demonstrated that stars can generate energy in sufficient quantity, through controlled nuclear fusion, to enable them to keep shining for millions to billions of years. How have stars managed to solve their problem of controlled fusion? As we shall now see, stars are able to do with ease what man is finding so difficult because they possess one enormous advantage. Because of their huge masses, stars can call upon gravity to act as the controlling agent.

To understand this role of gravity, let us use an imaginary episode from the life of Aladdin.

## THE GENIE AND THE SUN

The *Arabian Nights* story of Aladdin and the magic lamp ends with Aladdin living 'happily ever after' with his princess and his magic lamp. Here is a postscript to the story of interest to astronomers.

One hot summer's day, Aladdin, while on a tour of the Arabian desert, suffered sunstroke from which he took many days to recover. He was greatly annoyed with the Sun, and when he became well, he summoned the genie of the magic lamp and issued this command: 'Take the Sun apart and distribute its bits and pieces far and wide so that it is completely destroyed.' (See Figure 4-1.)

Now, assuming that the genie possesses boundless power and can undertake this mammoth task, just how much work is involved in the execution of Aladdin's command? Astronomers tell us that the Sun is nearly a spherical ball with a radius of nearly 700 000 kilometers and a mass* of approximately 2000 billion billion billion kilograms.

Of course, the genie soon realized that chipping off bits and pieces from the surface of the Sun and taking them far away does demand physical work. This is because each bit is attracted by the remainder according to the law of gravitation. To take any particular bit away from the rest, the genie has to work *against* this force of gravity. What is the total amount of work the genie would have to do to take the Sun completely apart and to move all its bits and pieces far away? The precise answer to this question will depend on the exact distribution of matter in the Sun. But the answer is of the order of $G \times M \times M/R$, where $M$ is the mass of the Sun and $R$ is its radius. $G$ is the constant of gravitation (which we encountered in Chapter 2 in the statement of Newton's law). With the values of $M$ and $R$ just given, this quantity of work turns out to be about $4 \times 10^{41}$ joules. We will shortly put this large quantity in proper perspective. For the time being, let us denote it by the symbol $W$.

Aladdin began to have second thoughts long before the genie completed the job. He realized how essential the Sun was to the inhabitants of the Earth, including himself. So, while the genie

* See Chapter 3 for a discussion of how the law of gravitation enables us to measure the mass of the Sun.

Figure 4-1. The genie of the magic lamp takes the Sun apart.

was in the process of executing the job, Aladdin issued his next command: 'Put all the bits of the Sun back together.' The poor genie went back to execute the second command.

However, this time, to bring all the constituents of the Sun together, the genie no longer had to work against gravity. In fact, while the genie had been temporarily called away by Aladdin to issue his second command, the bits and pieces left in space by the genie had already begun to fall back together. Gravity, which had been an opposing force for the first job, had now turned into an ally. And to put the Sun back together, the genie had to do no work. Instead, the amount of work $W$, which the genie had earlier expended on the first job *against* the force of gravity, would now be done by the force of gravity to put the Sun back together.

Let us now recall the concepts of work and energy outlined in Chapter 1. There we saw that, for a moving body, the work done *by* the impressed force results in an *increase* in the kinetic energy of the body. In the story of Aladdin, we note that gravity (as an impressed force) does work to bring the Sun to its present shape from an initially dispersed state. When we put these two ideas together we arrive at the *Kelvin–Helmholtz contraction hypothesis* for explaining why the Sun shines.

## THE KELVIN–HELMHOLTZ HYPOTHESIS

Two distinguished physicists of the last century, Lord Kelvin (1824–1907) and Baron Hermann von Helmholtz (1821–1894), proposed gravity as the primary source for stellar energy. Their hypothesis is called a *contraction hypothesis* because it states that the continued contraction of the Sun under its own gravity generates energy for radiation.

Consider the two states of the Sun shown in Figure 4-2. Stage I represents an early stage in the Sun's history. In Stage I, the Sun was much bigger than its present form, shown as Stage II. If the Sun is formed through condensation of an interstellar gas cloud, Stage I represents the state when the constituents of the Sun were well spread out. This is precisely the state the Sun was brought to by the genie under Aladdin's first command! From Stage I to Stage II, the Sun contracts under its force of gravity;

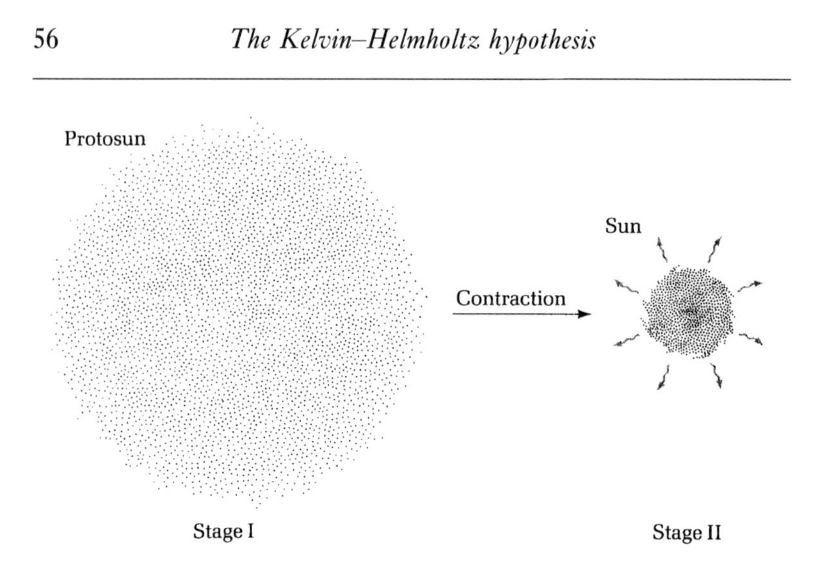


Figure 4-2. Two stages in the contraction of a gas cloud that becomes the Sun. Stage I is an early stage when the cloud is well dispersed and beginning to contract. In Stage II, the cloud has shrunk considerably because of its self-gravity. When the cloud has shrunk enough for nuclear reactions to begin, it radiates energy.

in other words, work is done *by* the impressed force of gravity in bringing the Sun to its present state from its primordial dispersed state.

By our rule for conversion of work to energy, this work by gravity should appear as kinetic energy, that is, the energy of motion. There is, however, no large-scale motion in the Sun. So where did this kinetic energy go?

If we were to examine this question carefully, we would discover that the kinetic energy has not disappeared! The Sun is in a gaseous state, and the particles of gas do move. As shown in Figure 4-3, the movement of gas particles is not systematic but *random*. Gas particles move in all possible directions with speeds ranging from small to large. Although these motions average out, leading to no systematic large-scale motion, the gas does have kinetic energy. And this energy increases (the gas particles move faster and faster) as the Sun slowly contracts.

If there is no manifestation of this kinetic energy in the form of a visible large-scale motion, how is the energy manifested? The effect of the kinetic energy is seen in two ways. One is through the rise

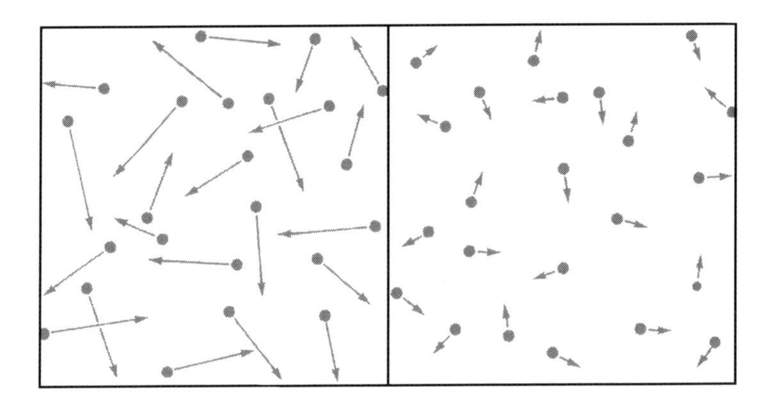

Figure 4-3. The random speeds of gas particles (shown above by large and small arrows) give rise to the property of pressure. This pressure will be exerted on the enclosure containing the gas by the impact of gas particles as they bounce against the walls and are reflected. The magnitude of the pressure is related to this impact, and in the figure above, it is larger on the left than on the right.

of *pressure* of the gas. As the kinetic energy rises, the gas pressure also rises.

Even on the surface of the Earth we talk of gas pressure when we describe the daily barometer reading. The statement 'the barometer reading is 760 mm' means that there is enough pressure in the atmosphere (see Figure 4-4) to support a vertical column of mercury 760 millimetres high. As we go up in an aircraft, the atmospheric pressure falls. At a height of 3000 metres, it becomes low enough to make it necessary to pressurize the aircraft.

The second effect of the change in the kinetic energy of a gas is manifested through a corresponding rise in temperature. Rise in the pressure is accompanied by a rise in the temperature of the gas. In the contracting gas cloud that eventually became the Sun, as the pressure increased, the temperature also increased. And gas at high temperature radiates light.

So, in the contraction hypothesis of Kelvin and Helmholtz, we have the following sequence of energy conversion:

$$\text{gravitational energy} \rightarrow \text{kinetic energy} \rightarrow \text{radiation energy.}$$

The Sun shines because of its gravity. Prima facie the argument seems reasonable.

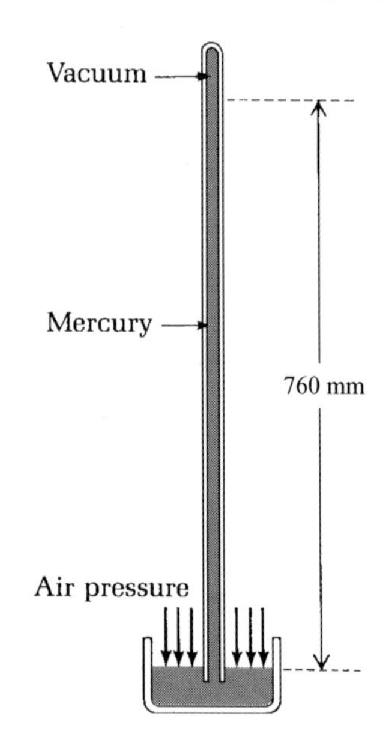


Figure 4-4. In the standard barometer, the pressure of the air (shown by arrows) is able to support the weight of a column of mercury. The height of the column gives a measure of the pressure. The part of the tube at the top is a vacuum. As the atmospheric pressure rises and falls, the column of mercury also rises and falls.

However, a scientific hypothesis must give a *quantitative* justification of what is observed. Simply saying that the Sun radiates light because of gravity is not enough. We need to know whether the gravitational energy reservoir of the Sun is quantitatively sufficient for all its radiation.

To check this, let us consider $W$, the amount of work done by the genie. $W$ is also the energy the Sun has expended during contraction from Stage I to Stage II. How long has the contraction gone on? To calculate this time interval, we need to know the rate at which the Sun has been expending this energy through radiation. From the amount of radiation received on the Earth, astronomers calculate this rate of energy expenditure to be about $1.2 \times 10^{34}$ joules per year. If this rate has not changed substantially over the Sun's lifetime to date, then gravitational energy has kept the Sun shining for a period of about *30 million years*.

Thirty million years is a long time span by human reckoning, and again our first reaction is that the Kelvin–Helmholtz hypothesis has provided a satisfactory explanation of solar luminosity. However, difficulties with this hypothesis surfaced when geologists estimated the age of the Earth to be considerably greater than 30 million years. The present estimate of the age of the Earth is nearly *4.6 billion years!* If the present ideas about the origin of the solar system are to be trusted, the Sun and the Earth must have formed at about the same time. And there is evidence to suggest that the Sun must have been shining at the present rate for a period of this order. If the Sun is considerably older than 30 million years, then the Kelvin–Helmhotz hypothesis is inadequate and we must look to some other source than gravity for an explanation of its energy reservoir.

## THE SUN AS A FUSION REACTOR

The mystery of the Sun's energy reservoir remained unsolved until the third decade of the present century. By then, astronomers had begun to have clearer ideas about the internal constitution of the Sun and other stars. The Cambridge astronomer Sir Arthur Eddington was able to express these ideas in the form of four equations of stellar structure.

These equations essentially summarize the following information. The first equation is called the *equation of hydrostatic equilibrium* (see Figure 4-5). It describes how the Sun (or a star) is held in equilibrium under the opposing forces of gravity and internal pressures. The internal pressure in a star partly arises from the hot gas in its interior and partly from the radiation emitted by the hot gas. Gravity has the tendency to shrink the Sun, whereas internal pressures tend to expand it. The second equation describes how the mass of the Sun is related to its density. The third equation, known as the *equation of state*, connects the pressure at any internal point to the ambient temperature and density. The overall effect of these equations is to generate a model of the Sun as a sphere of gas with a high temperature in the centre that progressively decreases outward. The fourth equation describes how the radiation generated in the hot inner regions is progres-

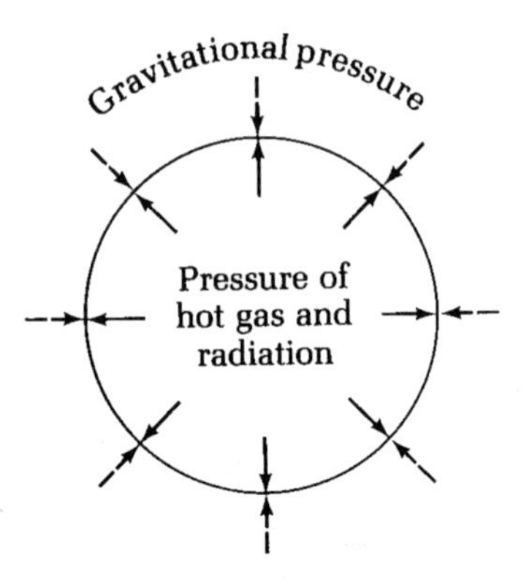


Figure 4-5. The opposing forces of gravity (inward dashed arrows) and pressure (outward solid arrows) are shown acting on any spherical surface inside and concentric with the surface of a star. For the hydrostatic equilibrium of the star, these forces must be in exact balance.

sively absorbed as it moves outward. Because of the absorption, only a small fraction of the inner radiation can escape from the surface.

By studying the radiation intensity at the Sun's surface for different wavelengths the astronomers are able to estimate the temperature at the solar surface. The answer comes out close to 5500 degrees Celsius. With his equations, Eddington was able to show that a viable model of the Sun with this surface temperature has a central temperature higher than 10 million degrees. In other words, hot though the Sun appears at its surface, deep down at its centre it is hotter by a factor of about 2000.

The missing piece of information at this stage was the mysterious energy reservoir of the Sun. Clearly, with radiation flowing out from the deep interior of the Sun there had to be a source for it in the central region. Here, Eddington made a prophetic suggestion. He argued that the central temperature in the Sun was high enough to release nuclear energy in sufficient quantities to provide for the Sun's luminosity.

However, Eddington's conjecture attracted opposition from the atomic physicists. They felt that stellar interiors are not hot enough to trigger release of nuclear energy. To such critics, Eddington gave the spirited reply,* 'We do not argue with the critic who urges that the stars are not hot enough for this process; we tell him to go and find a hotter place.' Hell hath no fury like a theoretician scorned!

In the 1920s, nuclear physics was a new subject, and neither Eddington nor his critics had enough knowledge to pursue the argument further. Eventually, Eddington was proved right – the temperatures in the central regions of stars *are* high enough to sustain nuclear reactions with the fusion of light atoms.

With our present knowledge of the nucleus, it is possible to understand why the controversy arose in the first place and how it was subsequently resolved. In Figure 4-6a, we see four separate hydrogen (H) nuclei (which are the positively charged *protons*) while in 4-6b we see the nucleus of a helium atom (He). This has two protons and two *neutrons*. In a fusion reaction, the four H nuclei are brought together and converted to a He nucleus:

$$4\mathrm{H} \rightarrow \mathrm{He} + 2\mathrm{e}^{+} + 2\nu + \text{ energy}.$$

This symbolic way of writing the reaction tells us that the products of the reaction are a helium nucleus, two positrons ($e^{+}$), two neutrinos ($\nu$), and release of energy. *Positrons*, the antimatter counterparts of *electrons*, have the same mass as electrons but are positively charged. Indeed, if we were to require that the total electrical charge remain unchanged in a fusion reaction, then it follows that, besides the He nucleus, two units of positive charge must be carried by other reaction products. Positrons do this job.

The release of energy in the above fusion reaction takes place for the following reason. The masses of the four participating H nuclei taken together exceed the sum of the masses of the reaction products (the He nucleus and the other four light particles) by a small amount. Einstein's special theory of relativity states that loss of mass in any process is balanced by a corresponding gain in energy. This energy is related to the loss of mass by the famous formula $E = Mc^2$.

In the fusion reaction leading to the formation of the He nucleus,

* See A. S. Eddington, *The Internal Constitution of the Stars*, Cambridge University Press, 1926, p. 301.

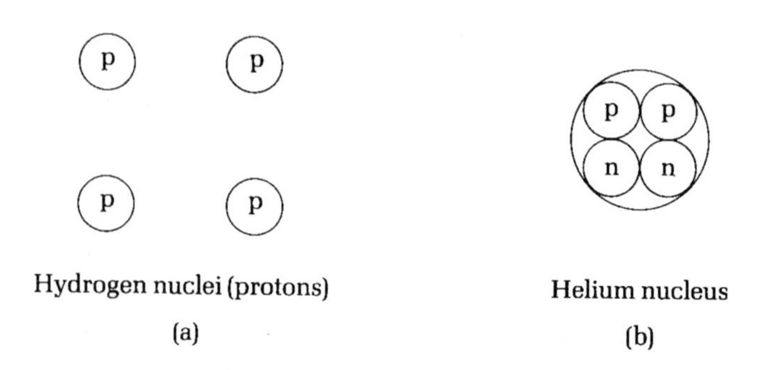


Figure 4-6. In (a) we have four hydrogen nuclei, which are in fact four positively charged protons (p). In (b) we see the helium nucleus with two protons and two electrically neutral particles, neutrons (n). In a star, the material in (a) is transformed to the material in (b). In this process some energy is released.

the mass that is lost is equivalent to the energy of 26.72 MeV*. To put it differently, of the hydrogen converted to helium, the fraction of mass that is converted to energy is 0.7%. This is the energy reservoir that we humans on Earth are trying to tap through our attempts to build a fusion reactor.

The reaction in the fusion reactor differs somewhat from fusion in the Sun. In the fusion reactor on Earth, the primary fuel is *heavy* hydrogen. Its nucleus has a neutron as well as a proton. *Two* such nuclei have to be fused to yield a He nucleus and radiant energy.

The atomic physicists of the 1920s objected to Eddington's hypothesis because of the difficulty of bringing four H nuclei together. Such protons are positively charged; they repel each other according to the electrostatic law that *like charges repel.* And the force of repulsion rises steeply as the charges come closer. How then could these like charges be brought together? The difficulty looked insurmountable in the 1920s but was resolved in the next decade with a better appreciation of the attractive nature of the strong nuclear force. Notice that the He nucleus of Figure 4-6b has two protons held together. How can this happen if two like charges repel each other? The answer is that, within the nucleus, a new force, much

* MeV, or million electron-volts, is an atomic unit of energy. One kilowatt hour, the unit commonly used for measuring electricity consumption, equals about $2 \times 10^{19}$ MeV.

stronger than the electrostatic force of repulsion, acts in such a way as to hold the four particles (the two neutrons and two protons) together. This strong nuclear force acts on neutrons as well as on protons, but it acts over a very short range. If protons can bombard each other with sufficient speeds, they can come close enough to feel the effect of the strong nuclear force. In a high-temperature hydrogen gas, the H nuclei would have large random motions and thus would occasionally surmount the electrostatic repulsion and come close enough to be fused by the strong nuclear force. The temperatures in the centres of stars, which range from 10 to 40 million degrees Celsius, are high enough to endow the H nuclei with great enough speeds to bring them together and thereby trigger nuclear fusion, as Eddington had argued.

## GRAVITY AS THE CONTROLLING AGENT

The modern theory of stellar structure is based on the four equations set up by Eddington, together with a fifth equation describing the rate of energy generation in the fusion reactions in the central core of the star. In 1938, Hans Bethe wrote down the missing fifth equation and constructed a complete model of the Sun.

The crucial role of gravity in these equations cannot be overemphasized. Huge pressures are needed inside the gaseous Sun in order to balance the attractive force of gravity and prevent total contraction. These pressures are linked to high temperatures and densities within the gas. Typically, a contracting cloud of interstellar gas becomes a star when the temperature in the core reaches a high enough value to trigger nuclear reactions.

In the terrestrial attempts to produce high temperatures suitable for triggering nuclear reactions, the above controlling effect of gravity is absent. In the deep interior of the Sun, gravity confines the gas undergoing explosive nuclear energy generation. On Earth, we have to look for other means, such as a magnetic field, to confine the hot gas. Devices known as *tokamaks* attempt to achieve this, but so far a working fusion reactor is far from reality.

Let us now make a thought experiment to further explore the effect of gravity on stars. Suppose we connect a hot star to a cold star by a conducting wire. We know the heat flows from a hot body to a cold body, and accordingly in our thought experiment heat will flow from the hot star to the cold star.

Nevertheless, there is a surprise in store for us! Under normal circumstances, when heat flows from the hot to the cold body, the temperature of the hot body is lowered while that of the cold body rises. In our example, when heat passes from the hot star, its internal pressures drop and its equilibrium is disturbed. Because its pressures are no longer strong enough, the star contracts under its force of gravity (see Figure 4-5). As the star contracts, its gas heats – thus the hot star becomes hotter! What happens to its cold companion? As it receives heat, its pressures rise and its equilibrium is also disturbed. This star *expands* as its internal pressures become stronger than its force of gravity. As the star expands, its gas cools further. So the cold star becomes colder!

Strange though this behaviour is, something like it does occur during the evolution of a star. We have already seen that, in the central core of a star like the Sun, the temperature is high enough to sustain the fusion of hydrogen into helium. What will happen when the hydrogen in the core is exhausted? The fusion reactor will be temporarily switched off for want of fuel. This will lead to a drop in the heat production and in the pressure in the core. The core therefore shrinks and heats up. As its temperature rises and reaches, say, 100 million degrees Celsius, the reactor comes alive again. However, now the fuel to be burnt is not hydrogen but helium. At this temperature, three helium nuclei can be fused to form a carbon nucleus, as shown in Figure 4-7. Meanwhile, to preserve overall equilibrium, the star's outer envelope *expands*. The star becomes a *giant star*. The expansion of the envelope leads to cooling, so that the surface temperature of the star drops. Whereas the Sun has a surface temperature of about 5500 °C, a giant star can have a surface temperature as low as 3500 °C. The colour of a giant star is therefore closer to red, in contrast to the predominantly yellow colour of the Sun (see Figure 4-8).

## STELLAR NUCLEOSYNTHESIS

Having seen Eddington's encounter with the atomic physicists (on the matter of hydrogen fusion) in the 1920s, we find that the next

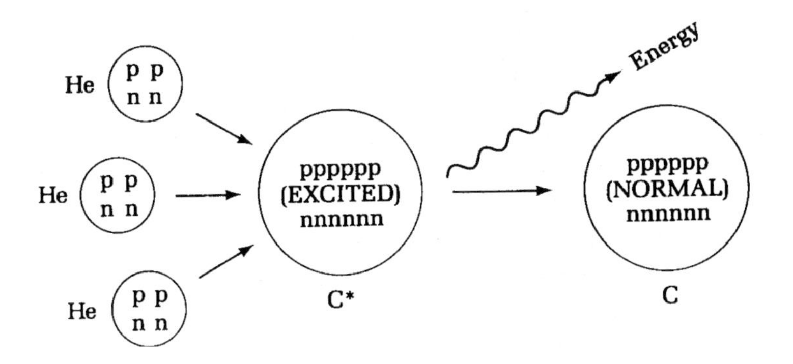


Figure 4-7. The diagram shows that three helium nuclei, each containing two protons and two neutrons, are fused through a resonant reaction into a nucleus of carbon. Fred Hoyle has pointed out that this nucleus is in an *excited* state and it decays into a normal carbon nucleus with the emission of energy. This nuclear reaction occurs in later stages of stellar evolution when a star's hydrogen fuel has been fused into helium nuclei. Helium fuses at a higher temperature than hydrogen.

fusion process (of conversion of helium to carbon) led in the 1950s to another encounter between a theoretical astrophysicist and nuclear physicists. This time the theoretician was Fred Hoyle, who was subsequently to occupy the Plumian Professorship at Cambridge held earlier by Eddington.

Hoyle was faced with two questions. First, what happens to a star like the Sun after it has exhausted all its hydrogen fuel? Can helium formed at the centre be used as a fuel to make still heavier nuclei? This would prolong the star's existence as a radiating ball of light.

Hoyle's second question concerned the formation of nuclei of all the chemical elements found in nature. How did carbon, oxygen, metals etc. come into existence? Did stellar interiors provide conditions suitable for making all of them in a step by step process of nuclear fusion?

The trouble lay in the fact that having formed helium the next immediate lot of nuclei turned out to be unstable. Thus if one sought to add one hydrogen nucleus to one helium nucleus, the resulting nucleus with five particles would break back into the original pieces. The same happened if one sought to bring two helium

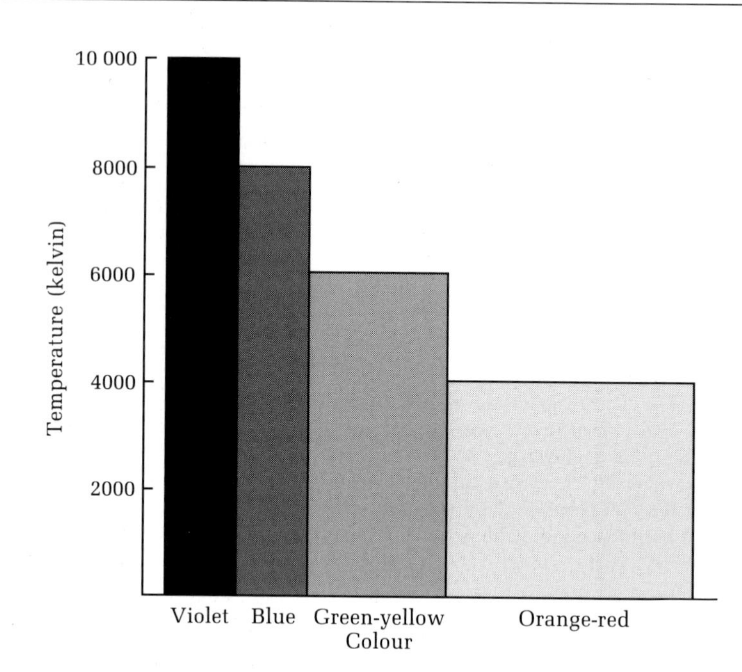


Figure 4-8. This figure shows the surface temperatures of stars in kelvin, which correspond roughly to the different colours in the visible part of the spectrum. The temperature increases as we go from red to violet. Thus a star with a temperature of around 8000 K will appear blue, whereas a star with a temperature of around 4000 K will appear mainly orange-red.

nuclei together. Thus the problem was like climbing a staircase which has steps 5 to 8 too weak to support human weight!

However, continuing with the staircase analogy, if one were bold enough to jump to step 12, one found it firm and stable. This is the carbon nucleus, which has six protons and six neutrons. But can one make such a jump? How can one bring three helium nuclei together?

This is where Hoyle came up with his remarkable suggestion. It was known that the coming together of three helium nuclei is indeed rare, too rare to make the fusion reaction work. But Hoyle found a loophole. If there were a resonant reaction, it would work fast, fast enough to compensate for the rarity of three helium nuclei coming closer.

In a resonant reaction the combined energy of the three He nuclei exactly matches the energy of a carbon nucleus. Just as in a musical string instrument the adjustment of tension in the strings causes the musical tones to resonate, so, with the above matching of energies, the reaction proceeds quickly. Hoyle found, however, that the normal carbon nucleus did not have enough energy to 'resonate'. So he predicted that the nucleus can exist in an *excited state* which has the appropriate higher energy.

He asked the nuclear physicists Ward Whaling and William Fowler at Caltech's Kellogg Radiation Laboratory to check whether such an excited state of the carbon nucleus indeed exists. They were skeptical at first, especially because of the astrophysical motivation of the conjecture. Eventually, however, they verified the conjecture and found it to be true.

This is the process described in Figure 4-7 and which solves both the questions bothering Hoyle. The three helium nuclei form into an excited carbon nucleus which subsequently decays to the normal state. In this process energy is released which keeps the star shining in a red giant state. Moreover, once carbon has been reached, our staircase can be climbed further! In this case each step amounts to adding an extra helium nucleus, thus forming a sequence of heavier nuclei:

$$\text{helium} \rightarrow \text{carbon} \rightarrow \text{oxygen} \rightarrow \text{neon} \rightarrow \ldots.$$

At each stage some fusion energy is available for the star so that it continues to shine. However, the process is a stop–go process. Once a particular fuel is exhausted, the inner core contracts and heats up, to a temperature at which the next reaction is triggered. At each stage in this process, the outer envelope of the star expands further to maintain equilibrium. The giant gets larger and larger. Nuclear physics tells us, however, that the process of fusion cannot just go on and on. In fact, it stops at the iron group of nuclei where the typical nucleus has 56 particles (photons and neutrons). Any further fusion of particles to the iron nucleus does not produce more energy. At this stage, the core temperature has reached around 10 billion degrees, and the star encounters a catastrophic situation. Gravity, which was so far able to exercise a restraining influence on the hot star, can no longer do so. Instabilities develop in the star, which results in the blowing away of its outer envelope.

Figure 4-9. A photograph of the Crab nebula. The bright object is believed to be the debris of a supernova explosion. (Courtesy of The Observatories of the Carnegie Institution of Washington.)

## STELLAR CATASTROPHES

This stellar catastrophe is seen as a *supernova* explosion. The photograph of the Crab nebula shown in Figure 4-9 is the most visually spectacular example of such a supernova explosion. The ancient records of Chinese and Japanese astronomers tell us that the explosion itself must have been observed from the Earth on 4 July 1054. The photograph shows the expanding outer envelope now, some nine centuries after the explosion.

The fallout of such an explosion takes the form of atomic nuclei (which have been fused inside the star), electrons, neutrinos, and radiation. The nuclei appear as showers of *cosmic rays*, which travel long distances in our Galaxy. It would indeed be catastrophic for us on the Earth if a supernova explosion occurred within a distance of, say, 100 light-years. The high-energy cosmic rays resulting from such an explosion would play havoc with the Earth's atmosphere. For example, they could strip it of all its protective layer of ozone, thus exposing life on Earth to the devastating ultraviolet rays from

Figure 4-10. Supernova 1987A: a photograph of the star which exploded is shown on the right and that of the post-explosion state on the left. (Courtesy of the Anglo-Australian Observatory, photograph by David Malin.)

the Sun. Fortunately, a supernova explosion is not very common. Throughout the Galaxy, the frequency of such explosions may be once in 100 to 300 years. So the chance of such an explosion occurring in our neighbourhood, up to a distance of 100 light-years, is as small as one part in a million per thousand years.

A spectacular sighting of a supernova explosion in modern times was on 23 February 1987 when the giant star known as Sanduleak in the Large Magellanic Cloud (LMC) was seen to explode. Supernovae are classified alphabetically each year in chronological order of their sighting. The SN 1987A was the first supernova to be seen in 1987. It was not in our Galaxy, but being in the LMC it was near enough to be spectacular. Figure 4-10 shows the photographs of SN 1987A before and after the explosion. It has been extensively monitored by various observational techniques and has provided valuable guidance for the theoretical ideas about such explosions.

Destructive though a supernova explosion appears to be, there is evidence that such an event may itself trigger star formation in a nearby gas cloud. The composition of our solar system suggests that its birth may have been triggered by such a supernova explosion. The shock waves generated from such explosions, impinging on an interstellar gas cloud, can set off its contraction. The Sun and the planets may have condensed out of such a contracting gas cloud. Thus stellar catastrophes can in this way play a constructive role as well as a destructive role.

What about the remnant of this explosion? What is left behind after the envelope has been cast off? We will return to this subject in Chapter 7.

# 5
# *Living in curved spacetime*

## IS NEWTON'S LAW PERFECT?

We left Chapter 2 with the impression that Newton's law of gravitation gave a successful account of the diverse nature of phenomena in which gravity is believed to play a leading role. Not only is this law able to account for motions of such celestial bodies as planets, comets, and satellites, it also helps us in understanding the complex problem of the structure and evolution of the Sun and other stars. Modern scientists use the same law in *determining* the rocket thrusts, spacecraft trajectories, and the timing of space encounters. That a good scientific law should be basically simple but universal in application is epitomized in Newton's law of gravitation. What more could one ask for?

Yet science by nature is perfectionist. The laws and theories of science are accepted as long as they are able to fulfil its primary purpose of explaining natural phenomena. Any law of science, despite a history of past successes, is inevitably discarded if it fails in even one particular instance. To the scientist, such an event brings mixed feelings. Disappointment and confusion that an old, well established idea has to be given up or modified are coupled with excitement and expectation that nature is about to reveal a new mystery.

Newton's law of gravitation was no exception to this rule. By the beginning of the present century, cracks were beginning to appear in the impressive facade of physics erected on the Newtonian ideas of motion and gravitation. The cracks were both conceptual as well as

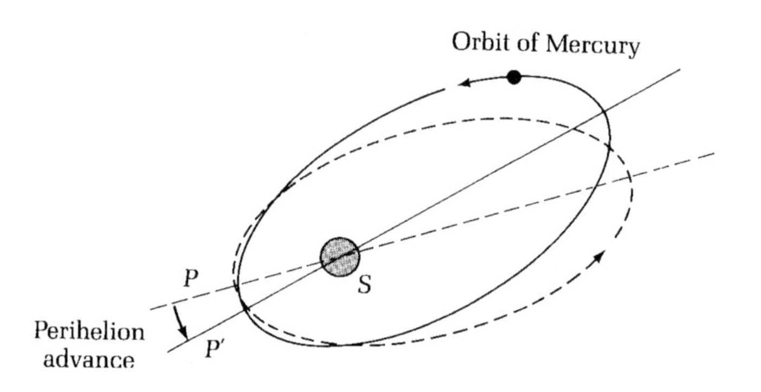


Figure 5-1. The planet Mercury (black spot) is expected to move in an elliptical orbit around the Sun ($S$), if Newton's laws hold. In practice the closest point ($P$) of the orbit from the Sun changes its position very slowly.

observational. It would take us too long to describe all the different issues involved, so we will limit ourselves to one example of each, taking the observational discrepancy first.

## THE STRANGE BEHAVIOUR OF THE PLANET MERCURY

The orbits of planets are supposed to be ellipses. This was Kepler's conclusion after a careful analysis of the data, a conclusion that was subsequently proved by Newton on the basis of his laws of motion and gravitation.

However, observations extending over several decades after 1764 began to reveal a minor discrepancy in the orbit of the planet Mercury. Of all the planets of the solar system, Mercury is the closest to the Sun and has the most eccentric orbit. It takes nearly 88 days to complete one orbit around the Sun. Thus, if Mercury is at the closest point $P$ to the Sun on a certain day, it is expected to return to that point after 88 days. This expected behaviour is shown in Figure 2-5.

Mercury, however, behaves anomalously in this respect. After completing the orbit, it does not return to the same spot. As shown in Figure 5-1, the point of closest approach to the Sun will have shifted from $P$ to $P'$.

The point of closest approach is called *perihelion*. The situation illustrated in Figure 5-1 implies that the perihelion of Mercury has advanced; and, as Mercury keeps going around the Sun, its perihelion will advance in the direction of the arrow shown in Figure 5-1.

The rate of advance is very slow. The line joining the Sun $S$ to the perihelion moves around in space so slowly that over 100 years it turns through an angle of only 575 seconds of arc. To put this in proper perspective, we should recall that an arc-minute is one-sixtieth of a degree, and an arc-second is one sixtieth of an arc-minute. Thus, in 100 years, the perihelion advance is only 9.58 arc-minutes, that is, less than a sixth of a degree.

Minor though this discrepancy is, its existence bothered the scientists. Recall that in Chapter 2 we encountered the case of a discrepancy in the orbit of the planet Uranus. There the apparent discrepancy was resolved when it was found that another (hitherto undiscovered) planet was perturbing the orbit of Uranus. Could the anomalous behaviour of Mercury be caused by a perturbing influence of other planets?

Calculations showed the answer to this question to be 'almost but not completely'. Of the total angle of 575 arc-seconds, a major part of about 532 arc-seconds is indeed due to the perturbing effect of other planets of the solar system. The residual effect of a perihelion advance rate of some 43 arc-seconds per century remained unaccounted for in the Newtonian framework.

This problem had become acute enough by the 1860s to worry astronomers. Leverrier who had played a key role in resolving the problem of Uranus in the 1840s was by now himself an established astronomer and he felt that the trick used in solving the earlier problem might work here too. So he proposed that the orbital discrepancy was due to a new intramercurial planet (which he even named *Vulcan*). However, the solution did not work this time; despite searches, no such planet was found. Thus the discrepancy between observations and the Newtonian picture remained unexplained.

Notice that the unexplained effect is less than 8% of the total effect and in absolute terms is extremely small. Nevertheless, to the perfectionist view of science, the surviving discrepancy did cast doubts on the validity of Newton's law of gravitation.

## FROM NEWTON TO EINSTEIN

Conceptual difficulties rather than observational discrepancies led Albert Einstein (1879–1955) to cast a critical look at the Newtonian law of gravitation. The most bothersome aspect of the law of gravitation was its concept of *instantaneous action at a distance*. The Sun and the Earth attract each other, according to this law, by a force that acts across the vast distance separating them – a distance of some 150 million kilometres. Not only is this force supposedly acting at great distances, it supposedly also acts *instantaneously*.

This instantaneous nature of gravity is illustrated by a thought experiment, for which once again we call upon Aladdin and his genie. Recall the incident in the previous chapter where Aladdin commanded the genie to take apart the Sun. Suppose instead that he had told the genie to annihilate the Sun at once by magic and without leaving a trace. Now, in the real world governed by the laws of physics, there are limitations on what the genie can do. For example, the law of conservation of matter and energy tells us that even the most powerful genie in the world cannot annihilate the Sun 'without leaving a trace'. If matter is destroyed, an equivalent amount of energy (given by the Einstein relation $E = Mc^2$ encountered in Chapter 4) must appear. But in the fantasy world of the *Arabian Nights* everything is permitted, and so let us continue our speculation on what would happen if Aladdin's command were carried out.

As shown in Figure 5-2, the moment the Sun disappears, the Earth will feel itself free from the gravitational pull toward the Sun. It will consequently take off in the direction tangential to its regular elliptical orbit. To the people on the Earth living on the part facing the Sun, the Sun's disappearance will be a visible fact (the day turning into night) some eight minutes *after* the Sun is destroyed, because light takes nearly eight minutes to travel from the Sun to the Earth. Thus, here we have a situation where an observable effect (the disappearance of the Sun) is communicated to the Earth by gravity faster than it is communicated by light.

Such a situation is contrary to the *special theory of relativity* developed by Einstein in 1905. This theory places an upper limit on the speed with which an observable effect can propagate from one

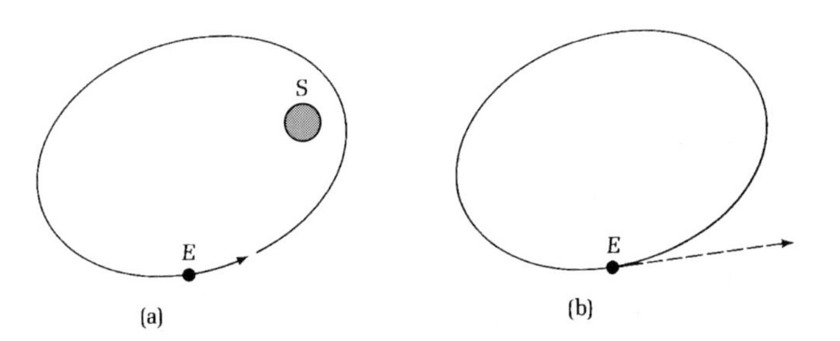


Figure 5-2. If the Sun were to disappear by magic, the Earth, freed from the Sun's gravity, would take off in the tangential direction to its elliptical orbit. (From *The Physics–Astronomy Frontier* by F. Hoyle and J. V. Narlikar. Copyright © 1980. W. H. Freeman and Company.)

point in space to another. *This upper limit is represented by the speed of light.* Thus the notion of instantaneous propagation of gravitational pull across vast distances is inconsistent with the basic tenet of special relativity. That is why, having developed the special theory of relativity, Einstein was compelled to revise the law of gravitation.

At this stage, the reader may ask, 'What is this special theory of relativity? Why is it so important that to conform with it we must change the law of gravitation so well established over two centuries?' In this book, we are concerned mainly with the subject of gravity and hence cannot devote much space to these interesting and important questions. The following brief discussion, while sufficient for our purpose, hardly does justice to the important place that special relativity occupies in modern physics. The revolution brought about by this theory in modern physics is comparable to the revolution brought about by Galileo in the medieval physics dominated by Aristotle.

Special relativity questioned the validity of the fundamental concepts of *absolute space* and *absolute time*. In Figure 5-3a, we see two points, $P$, $Q$, in space. If we are asked to find the distance between $P$ and $Q$, we would place a ruler along the line $PQ$ and measure the length of the segment. In Figure 5-3b, we show two instants of time, $A$ and $B$, at the same point in space. To measure the period elapsed between $A$ and $B$, we use a clock. The number of ticks of the clock between $A$ and $B$ gives us the measure of the time interval between $A$ and $B$. Now, our intuitive feeling is that these measure-

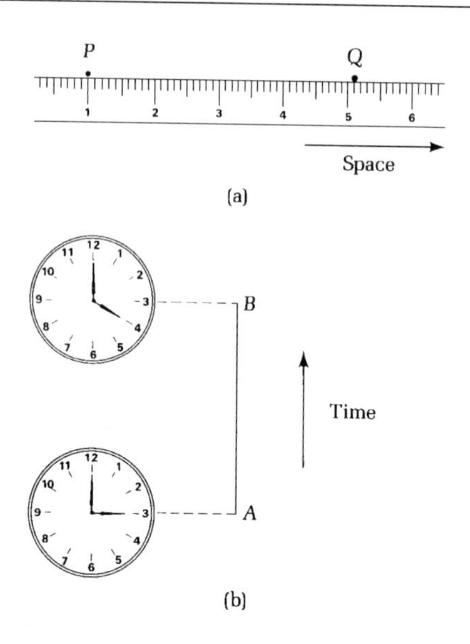


Figure 5-3. The measurements of spatial displacements (a) and time intervals (b) in Newtonian/Galilean dynamics.

ments have an *absolute* character, that is, they do not depend on the observer. In particular, if two observers moving with different speeds make these measurements, they should come up with the same answer. This is the intuitive feeling on which the physics of Galileo and Newton was based. This was the notion that Einstein challenged.

Of course, concepts in science may have their origin in the intuitions of scientific geniuses, but their ultimate validity rests on physical experiments. Recall from Chapter 2 that Galileo himself emphasized the role of experiments in his reasoning.

The concepts of absolute space and absolute time seemed to rest on solid foundations. Yet, toward the end of the nineteenth century, with improvements in the accuracy of laboratory experiments, cracks began to appear in these foundations. One experiment that played a very important role in this context was the experiment of Michelson and Morley in 1887. The experiment, which is described in Figure 5-4, showed that, in an interferometer, the time taken by light to make a return journey in the east–west direction is

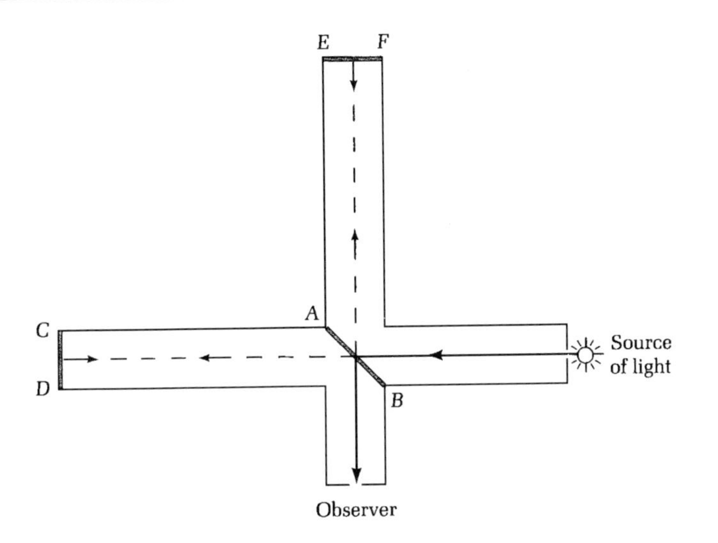


Figure 5-4. In the Michelson–Morley experiment, a ray of light (from a source on the right) falls on a partially reflecting and partially transmitting mirror *AB* in Michelson's interferometer. The reflected part goes up and is reflected by the mirror *EF*. The transmitted part goes in the original direction and is reflected by the mirror *CD*. The two parts recombine and are seen by the observer. If we take into account the fact that light is a wave, the result of the above recombination depends critically on the phases of the two split waves. In the extreme case of the crests of the two waves falling together, the total light doubles, while in the opposite extreme the crest of one wave can cancel the trough of another. In general, a series of dark and bright fringes are seen by the observer. The interference of two waves depends on how far each wave has travelled as well as on the speed of light. Since the two arms of the interferometer are equal in length, the shifts in the interference fringes could be used to detect minute changes in the speed of light. Michelson and Morley used this technique to measure the expected difference of light travel time in the north–south and east–west directions. They failed to find any difference.

the same as the time taken for a return journey of the same length in the north–south direction. If Newtonian ideas of motion were correct, the answer should have been different in the two cases because the experiment was performed on the Earth's surface, which moves from west to east. This motion should have led to light taking a longer time in the return journey along the east–west route than along the north–south one.

The experiment was accurate enough to detect the effect of (west-to-east) motion of the Earth's surface as it rotates relative to 'absolute space'. The surprising result that no effect was detected led to considerable speculation among the leading physicists of the nineteenth century.

Einstein's work in 1905 correctly resolved the implications of the Michelson–Morley experiment. Einstein had set out to re-examine the concepts of absolute space and absolute time in terms of James Clerk Maxwell's laws of electricity and magnetism, formulated in the 1860s. He had come to the conclusion that, in order to preserve the inherent symmetries of Maxwell's laws, it was necessary to revise the Newtonian concepts of absolute space and absolute time. Einstein interpreted the Michelson–Morley experiment in the most direct and straightforward terms – that the speed of light was unaffected by the motion of the Earth. In fact, this was a special case of Einstein's more general conclusion that the speed of light is the same for all observers moving relative to each other.

The first law of motion, originally discovered by Galileo, singles out a special class of observers – those that move with a uniform velocity, that is, those observers on which no force acts. Such observers are called *inertial observers*; any two inertial observers move with uniform velocity relative to each other. According to Einstein, the speed of light will be the same as measured by them both. This conclusion is a special case of the more general statement that the basic laws of physics are the same for all inertial observers.

If we observe a passing express train from the platform of a railway station, it appears to flash past. If we observe the same train from a fast-moving car on a parallel highway going in the same direction, the train does not appear to move as fast. This is because the velocity of the train relative to the car is much less than it is relative to the stationary platform.

Contrast this result with Einstein's conclusion that the speed of light is the same with respect to all moving inertial observers and you begin to see why such a conclusion is against our intuition. But, as the Michelson–Morley experiment so conclusively demonstrated, nature does not always respect our intuition!

Faced with this startling conclusion, Einstein had to revise the concepts of measurements of spatial distance and time intervals. The notions of absolute space and absolute time, as illustrated in Figure 5-3, were inconsistent with the constancy of the speed of light. Instead, space and time measurements became *dependent* on the inertial observer making the measurements. Two observers in relative motion making measurements of the type shown in Figure 5-3 *will not get the same answer*. The rules connecting their measurements are known as the *Lorentz transformation*, after the physicist Hendrik Anton Lorentz*. Thus, in special relativity, the Galilean concepts of absolute space and absolute time were replaced by a unified concept of space and time in which neither space nor time by itself had an absolute status. The laws of motion written according to the new rules of the Lorentz transformation implied that several changes needed to be made in the concepts of Newtonian dynamics. In particular, the new rules implied that the speed of light represents an upper limit on the speed at which any material particle or physical information can be transmitted.

Against this background, in the years 1905 to 1915, Einstein looked for a theory of gravitation that possessed all the successful features of the Newtonian law of gravitation and yet was free from the conflicts between the Newtonian law and the special theory of relativity.

## THE GENERAL THEORY OF RELATIVITY

One of the difficulties that threatened the coexistence of the Newtonian law of gravitation with special relativity has already been mentioned. Special relativity attached a special significance to the speed of light as the upper limit that cannot be exceeded by any physical interaction, whereas gravity (according to Newton) seemed to be operating instantaneously across vast distances. An inertial observer is one on whom no force acts. Can we actually pinpoint any observer or any physical object that is free from all forces? In our discussion of the first law of motion, we encountered friction as a force that retards motion. However, under idealized conditions, we can achieve situations where the friction is made very, very small. A man attempting to walk on an ice rink knows how difficult it is

* Lorentz had given these rules in connection with another theory in his own attempts to explain the result of the Michelson–Morley experiment. Einstein found these rules applicable to his special relativity, although in a different context.

Figure 5-5. Einstein managed to fit a square peg into a round hole by modifying both the peg and the hole! His general theory of relativity resolved conflicts between Newton's theory of gravity and the special theory of relativity.

to proceed with very little friction. A projectile fired in a vacuum chamber encounters no resistance from air. In both these cases, however, there is another force that has been ignored: the force of the Earth's gravity. Gravity acts on all material objects and can never be eliminated as a force. Even if we go far, far away from the Earth, we still have the gravity of other objects in the Universe to contend with.

In short, there is nowhere in the Universe that we can go to eliminate gravity as a force, and so our definition of the inertial observer seems unrealizable in practice. Since inertial observers form the starting point of special relativity, it looks as if the theory is based on unrealistic foundations. Thus, not only does special relativity make the Newtonian law of gravitation inconsistent, but its own foundations appear threatened by the existence of gravity.

Einstein discovered an ingenious way out of these difficulties by proposing an entirely new approach to the phenomenon of gravity (Figure 5-5). His theory of gravitation, called the *general theory of relativity*, is not a patchwork repair job on Newtonian gravitation and special relativity; rather, it is a radically new attempt at understanding the nature of gravity and motion together. Let us try to understand the reasoning that leads to this remarkable theory.

We have already seen how all-pervasive gravity is and the futility of trying to ignore its existence anywhere in the Universe. Einstein took this property of gravity as evidence that it is intimately linked with another all-pervasive entity around us, *spacetime*, that is, the three-dimensional space in which we operate and time whose passage we all experience. For the linking agent between gravity and spacetime, Einstein proposed *geometry*.

Geometry is a branch of mathematics that is devoted to the study of measurements of angles and lengths of various shapes and figures. The rules of geometry were first systematically stated by the Greek mathematician Euclid (*c.* 300 BC). Starting from certain apparently reasonable axioms, Euclid developed theorems about triangles, squares, circles, and other figures. For a long time, mathematicians believed Euclid's axioms to be absolutely true and incontrovertible, and Euclid's geometry acquired a unique status as *the* system describing any measurements made in space.

However, axioms are assumptions whose validity cannot be proved. Any set of axioms that are logically self-consistent can form an independent branch of mathematics. Euclid's geometry is merely one example of many possible self-consistent sets of axioms.

For example, one of Euclid's axioms relates to the existence of parallel lines. In Figure 5-6, we see a straight line $AB$ and a point $P$ outside it. Euclid's axiom asserts that one and only one line can be drawn through $P$ parallel to $AB$. The line $CD$ shown in Figure 5-6 is parallel to $AB$ – that is, $AB$ and $CD$ will not meet even if they are extended in both directions. To our intuition, this assumption looks reasonable and *true*. But it is, nevertheless, an assumption. It cannot be proved on the basis of other axioms. Indeed, many mathematicians mistakenly thought that this axiom *could be proved*, and some even offered what eventually turned out to be erroneous proofs. Only in the last century was it finally realized that Euclid's parallel axiom is an assumption that cannot be proved. Moreover, mathematicians were able to show that, by altering Euclid's parallel axiom, new geometries could be created that were internally self-consistent. The axiom could be altered in two ways – either by asserting that through $P$ *no line* can be drawn parallel to $AB$ or that through $P$ *more than one line* can be drawn parallel to $AB$. The work of mathematicians such as Lobatchevsky (1793–1856), Bolyai (1775–1856), Gauss (1777–1855), Riemann (1826–1866), and others led to whole new geometries based on an alteration of the parallel axiom. These geometries came to be known as *non-Euclidean geometries*.

Let us now return to Einstein and his attempt to link gravity to

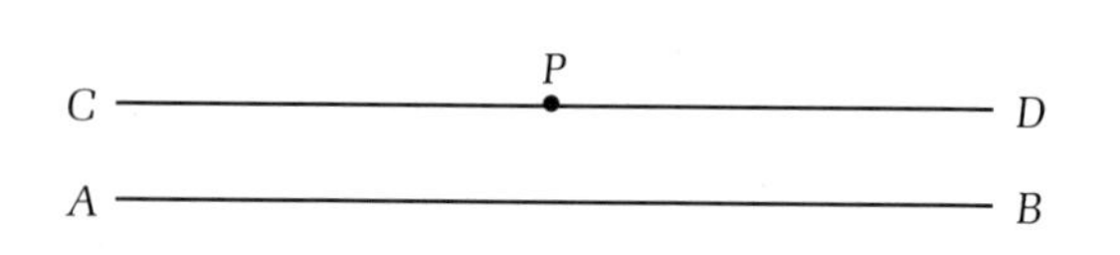


Figure 5-6. In Euclid's geometry, the parallel axiom states that through $P$ one and only one straight line can be drawn parallel to $AB$. The line $CD$ is such a line.

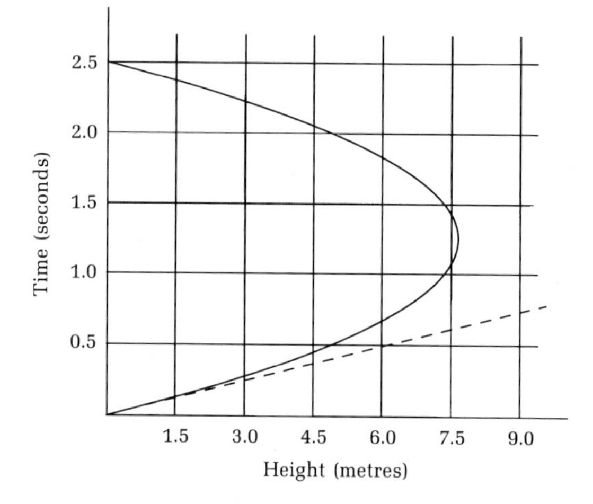


Figure 5-7. The world line of the ball is shown by the solid curve. The dashed straight line shows what this world line would have been in the absence of gravity.

spacetime. To illustrate Einstein's reasoning, we will revive our example of the ball thrower in Chapter 3. In Figure 5-7, we illustrate the trajectory of the ball in a spacetime diagram. Recall that the ball was thrown up with an initial speed of 12 metres per second, reached a height of 7.5 metres, and then fell back. The trajectory shown as a continuous curve in Figure 5-7 tells us where the ball is to be found at any given instant during its flight. Such a trajectory is called the *world line* of the ball. The world line in Figure 5-7 is a curve known as a *parabola*.

Let us now speculate what would have happened to the ball *if there were no gravity*. By the first law of motion, the ball would have continued moving with the same speed (12 metres per second) in the upward direction. Its world line then would be shown by the dashed straight line in Figure 5-7.

A Newtonian scientist would say that, in the absence of gravity, the straight line describes the ball's trajectory, whereas with gravity the straight line is bent to the parabolic shape. To this statement, Einstein would have replied that there is no such thing as the absence of gravity, and hence the dashed line in Figure 5-7 has no real status. The only line of physical significance is the curved line. How should one interpret the curved line by itself?

Einstein's reply to this question was that this line in fact describes the motion of the ball *under no forces* but in a spacetime whose geometry is changed by gravity. Because of the Earth's gravity, the spacetime has acquired a *non-Euclidean geometry*, and in that geometry the ball's trajectory is straight.

This may appear crazy. You might be tempted to say that the continuous curve in Figure 5-7 is obviously not straight, but let us ponder Einstein's reply. How do we define a straight line? It is the line of shortest distance between two points. The shortest distance between two points is different in a non-Euclidean geometry than in Euclid's geometry. So what may be a straight line in one geometry may not be a straight line in the other, and vice versa. Thus, provided we know the exact quantitative nature of the non-Euclidean geometry in the spacetime around the Earth, we can test the veracity of Einstein's claim that the continuous line of Figure 5-7 is straight. How do we find out the details of this geometry?

Einstein's general theory of relativity gives us the method for determining this geometry. The theory tells us, in quantitative details, how the geometry of spacetime is related to the distribution of matter and energy. Einstein's equations of relativity are (like Newton's) compact and elegant to write down, but *in practice* these equations have turned out to be very complex. Thus the question of determining the spacetime geometry is not a simple one. Later we will discuss one solution that has proved immensely useful in testing Einstein's theory.

## NON-EUCLIDEAN GEOMETRIES

Before we consider observable implications of Einstein's theory, let us become familiar with the somewhat strange concept of non-Euclidean geometries. To begin, let us consider the surface of the

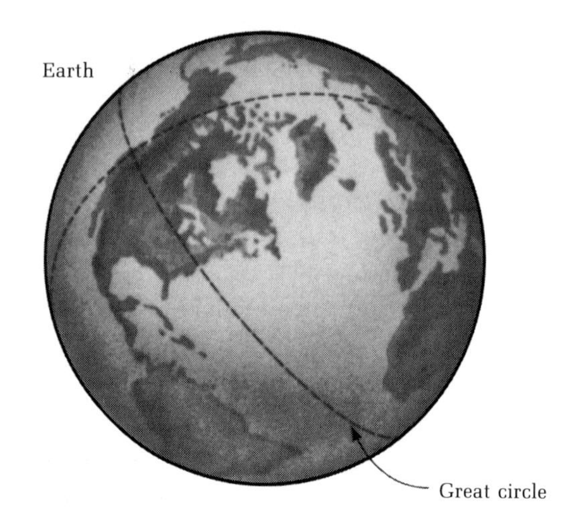


Figure 5-8. Straight lines on the surface of the Earth are arcs of great circles, that is, circles whose planes pass through the centre of the Earth. Notice that any two great circles meet on the Earth's surface.

Earth, idealized to be a perfect sphere. Imagine that the inhabitants of the Earth are flat, two-dimensional creatures sliding on the surface, with no appreciation of the third dimension of height. How will these creatures determine straight lines? Applying the prescription of shortest distance, two such creatures will stretch a piece of string between two points so that it is taut and lies on the surface of the Earth (see Figure 5-8). The resulting line is a segment of the great circle through the two points.

What is a *great circle*? If you slice a sphere with a plane saw passing through its centre, the rim of either piece will be a great circle. Thus the meridian circles on the Earth are great circles but the latitude circles (with the exception of the equator!) are not.

This is precisely the prescription that pilots would use to determine the shortest flight path between two airports. And this flight path will be different from the one got by using a ruler to draw a straight line on a flat map on paper. This illustrates once again our earlier remark that what constitutes a straight line depends very much on the basic rules of geometry. The geometry on the flat map is Euclid's, whereas that on the real Earth is non-Euclidean. For determining the path of the aircraft, it is the non-Euclidean geometry that is relevant.

There is another way in which we may think of a line as being straight. As we move along a curve, if we find that our direction of motion does not change, then we say that we are moving along a straight line. In Euclid's geometry, the straight line defined by the 'shortest distance' criterion also has this property. In a non-Euclidean geometry, these two criteria for determining whether a line is straight may differ from each other. However, in the type of non-Euclidean geometry chosen by Einstein, known as *Riemannian geometry*, these two criteria give the same answer.

Is there a simple method by which we can test whether the geometry on the surface of the Earth is non-Euclidean? Refer to Figure 5-8 and notice that *any two straight lines* (that is, great circles) *on the surface of the Earth intersect.* Thus Euclid's parallel axiom cannot be satisfied – no straight line can be drawn through $P$ parallel to $AB$.

There are other ways we can test the non-Euclidean character of the Earth's surface geometry. In Figure 5-9, we have a triangle drawn on the surface of the Earth that describes a journey undertaken by a flat creature starting from the North Pole, $N$. The journey proceeds in a straight line – which happens to be a meridian line – down to the equator. There the creature turns left and describes the segment $AB$ equal to a quarter of the equatorial circumference of the Earth. At $B$ the creature turns left again and returns to $N$ via the meridian $BN$. At $N$ he finds that his direction of return is at right angles to his starting direction. Now this triangle $NAB$ has a right angle at each of its three vertices, whereas the three angles of a Euclidean triangle add up to only 180 degrees. This example brings us to the notion of a *curved space.*

In the language of geometry, the two-dimensional space of the Earth's surface is *curved* and has *positive curvature.* The two-dimensional space on a flat piece of paper is *flat* or of *zero curvature.* The space on the surface of a saddle is also *curved* but with *negative curvature.* These curved surfaces are illustrated in Figure 5-10. Notice that, for a triangle drawn on a space of negative cur-

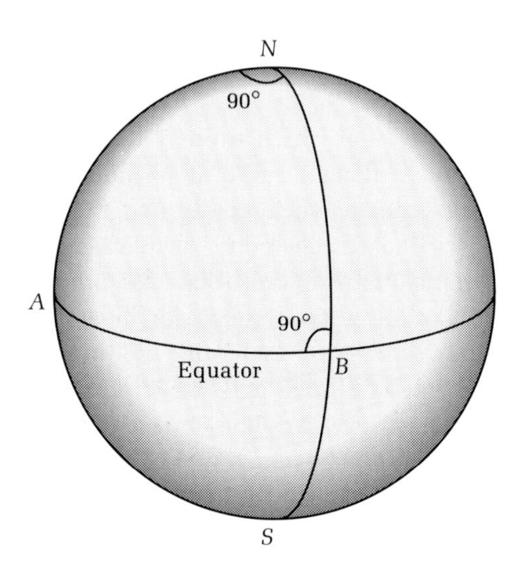


Figure 5-9. The triangle $NAB$ has three right angles.

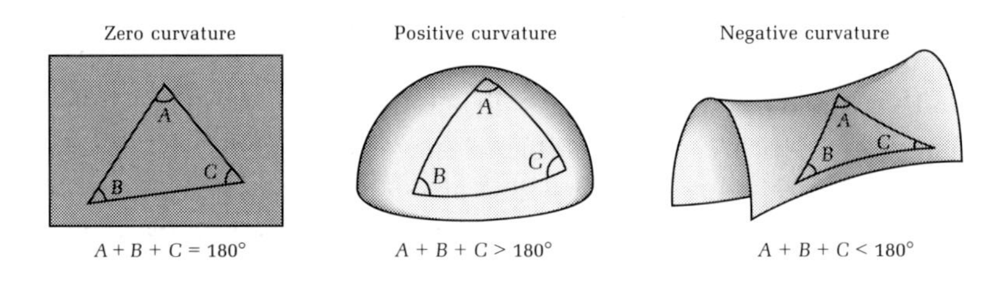


Figure 5-10. Examples of two-dimensional surfaces of zero, positive, and negative curvature. The three angles $A, B, C$ of a triangle sum differently in the three cases.

vature, its three angles add up to *less than* 180 degrees. Another way to decide whether a two-dimensional surface has a positive, negative, or zero curvature is to stretch a piece of paper on it and attempt to cover it point by point. If the paper covers the surface exactly, the surface has zero curvature; if the paper gets wrinkled, the surface has positive curvature; if the paper is torn, the surface has negative curvature. Try these criteria on various curved surfaces.

## DO WE LIVE IN A CURVED SPACETIME?

We now return to Einstein's interpretation of gravitation and ask whether there is any observational evidence that we live in a curved spacetime.

In 1916, soon after Einstein proposed his theory of relativity, Karl Schwarzschild (1873–1916) solved Einstein's equations to find out how the geometry of spacetime behaves if there is a massive spherical object in it. The *Schwarzschild solution* is the analogue of the Newtonian solution to the problem of how a spherical mass attracts other bodies gravitationally. This solution can therefore be used to determine, for example, how planets move around the Sun.

Recall Einstein's interpretation of the first law of motion in the presence of gravity. A planet's orbit around the Sun will be determined, according to this interpretation, by the criterion that it is a 'straight line' in the spacetime whose geometry is given by Schwarzschild's solution. And here we discover that despite different approaches there is considerable agreement between the final answers of Newton and Einstein. For all practical purposes, the planetary orbits as determined by Einstein's criterion are the same as those in the Newtonian theory! There are very slight differences, which are most noticeable for the planet Mercury because it lies closest to the Sun and has the most eccentric of all planetary orbits. Einstein's theory predicts that Mercury's orbit slowly rotates in space at the rate of about 43 arc-seconds per century.

Recall that in our earlier discussion, we did come across such behaviour on the part of the planet Mercury. We had found that the perihelion of its orbit advances at the rate of about 43 arc-seconds per century – an effect that had remained unexplained by the Newtonian law. Now we see that Einstein's theory provides an explanation. In fact, its prediction agrees extremely well with the observed rate of advance.

This remarkable success of general relativity inspires confidence in the theory. It was, however, another astronomical observation that established the viability of the theory. This was the observation of the bending of light rays near the Sun.

Light, we know, travels in a straight line. If the definition of

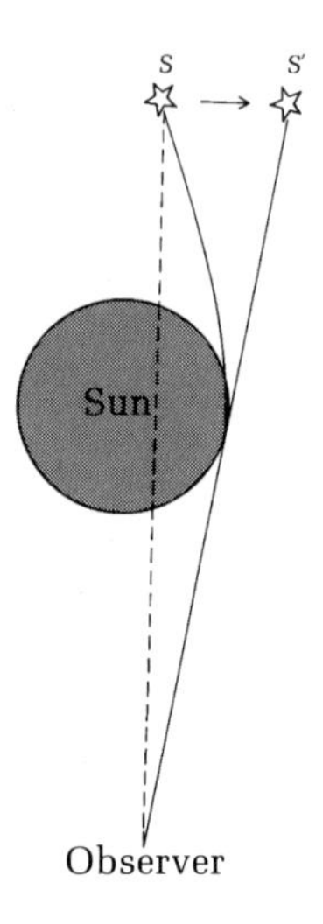


Figure 5-11. The bending of light by the Sun. Although the curved line that describes the light trajectory appears bent, it is straight according to Schwarzschild's geometry. The star $S$ appears to shift as the Sun comes close to its line of sight; its image is seen at $S'$.

a straight line changes according to the system of geometry, the path of a light ray will be different in curved space than in flat space. In Figure 5-11, we see two light paths passing close by the surface of the Sun. The dashed straight line describes the path in Euclid's geometry; the curved line is that path followed by light in Schwarzschild's geometry. To us, accustomed to Euclid's geometry, this latter path appears bent. But it is in fact the path of a light ray that is *straight* in the non-Euclidean geometry of Schwarzschild's solution. Which path does light actually follow?

Astronomers can answer this question by the following experiment. Suppose they observe a star whose line of sight is crossed by the Sun. In the situation where the Sun is away from the line of sight to the star, the difference between Schwarzschild's geometry and Euclid's geometry is negligible, and the Euclidean straight line from the star to us represents the path of light from the star. In Figure 5-11, just before (or after) the approach by the Sun, the light ray from the star $S$ grazes the solar limb and should show the maximum bending, as indicated by the curved line. The direction of the star should therefore appear to change as it is crossed by the Sun. The expected change in the direction, according to Einstein's theory, is very small – about 1.75 arc-seconds.

What prediction can one make on the basis of Newton's law of gravitation? Newton's law in its original form did not envisage that light would be attracted by the Sun at all, and so the expected change of direction is zero. However, if one assumes that light is made of tiny particles (called *photons*) that are subject to Newton's law of gravitation in the same way that particles of matter are, then the change in the star's direction in the above experiment should be exactly half that predicted by Einstein's theory.

Here then is an experiment that can, in principle, tell us which interpretation – Newton's or Einstein's – is right or whether both are wrong. In practice there are several difficulties. First, we cannot see the star when the Sun is shining brightly in the Sky. The only occasion when we can perform this experiment is when there is a total solar eclipse, not a very common phenomenon. Next, the expected change of direction is very small and requires very accurate measurement techniques. The situation is further complicated by the fact that, very near the Sun's surface, the hot gas can also bend the light rays just as a lens bends them.

As far as the first difficulty is concerned, astronomers did not have to wait long. In 1919, just four years after general relativity was proposed and only three years after the Schwarzschild solution, a total solar eclipse did take place. Realizing the importance of the experiment, the Cambridge astronomer Eddington took the initiative in making the measurements. Eddington and Cottingham went to Principe, an island in the Gulf of Guinea, while their colleagues Davidson and Crommelin went to Sobral in Brazil, places where the total eclipse could be seen. Thanks to a financial grant of £1000 obtained by the Astronomer Royal Sir Frank Dyson, these eclipse expeditions became possible.

It is interesting to note that in 1918, in the closing years of World War I, a Military Tribunal had discussed Eddington's appeal to be exempted from conscription on the grounds of being a 'conscientious objector' to war. At that time the argument that weighed strongly with the tribunal was that Eddington was planning the 1919 expedition for testing Einstein's theory of relativity and as such his presence and participation in the operations were essential. Eddington was allowed to continue his astronomical researches and could successfully execute the eclipse experiment.

The results of these observations favoured Einstein's theory rather than Newton's. When Sir Frank Dyson announced these results to a crowded meeting of the Royal Society in London, they caused a considerable sensation. Later, A. N. Whitehead recaptured the scene in the following words: 'The whole atmosphere of tense interest was exactly that of a Greek drama: we were the chorus commenting on the decree of destiny as disclosed in the development of a supreme incident. There was...in the background the picture of Newton to remind us that the greatest of scientific generalizations was now, after more than two centuries, to receive its first modification...'.

In retrospect we now see that, although the 1919 eclipse observations did go a long way toward establishing general relativity as a viable theory of gravitation, those observations were by no means conclusive. When an astronomer (or indeed any scientist) makes a measurement, his result is always subject to a number of imponderable or uncontrollable variations usually denoted as 'experimental errors'. Only when the errors are sufficiently small can one confidently assert that the reported measurement has confirmed or disproved a theoretical prediction. The range of errors of the Principe and Sobral eclipse observations was fairly large and though they were consistent with Einstein's predictions more accurate measurements were needed to substantiate them further.

Present technology makes it possible to do this experiment much more accurately using microwaves rather than visible light. Instead of a star, the quasar 3C 279 (3C here stands for the *Third Cambridge Catalogue*) is observed in the microwave region of the spectrum as it is occulted by the Sun. In this case, the Sun's own radiation is so small that we do not have to wait for an eclipse. Also, the solar atmosphere does not distort the path of microwaves as much as it distorts the path of visible light. The measurements made in 1975–1976 by the astronomers at the National Radio Astronomy Observatory in Green Bank, West Virginia, were so accurate that it can now be confidently asserted that Einstein was right and Newton wrong.

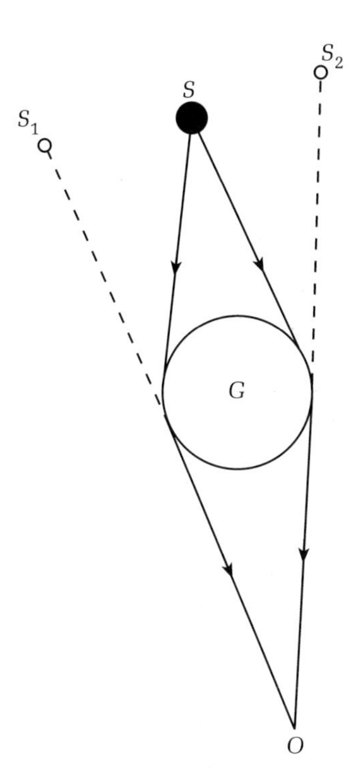


Figure 5-12. Two different paths followed by light from source $S$ to observer $O$, both deflected by a gravitating source $G$. $O$, therefore, sees two images $S_1$ and $S_2$ instead of the single source $S$.

## COSMIC ILLUSIONS

The example of a star's image shifting because of the bending of its light by the Sun's gravitation conjures up many more interesting possibilities.

Imagine, for example, a source of light $S$ and an observer $O$ with a massive object $G$ lying in between them. As happened in Figure 5-11, here too we have rays of light from $S$ to $O$ bent by the gravity of $G$. Figure 5-12 shows a possibility where there are *two* possible and different routes for light from $S$ to arrive at $O$. The light path is 'bent' by $G$ in both cases, but the interesting result is that both alternative routes are made possible by the non-Euclidean geometry in the neighbourhood of $G$.

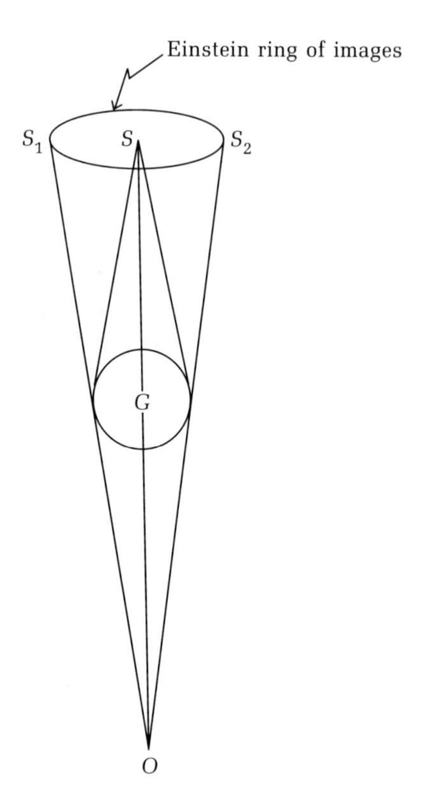


Figure 5-13. In a highly symmetric situation the images of the source $S$ are distributed along a ring perpendicular to the line of sight from $O$ to $S$.

Why is it interesting? Because the observer $O$ will now see not one but two displaced images of $S$ (shown by $S_1$ and $S_2$ in the figure)! And, of course, with more alternative light paths possible we might see even more images. In an idealized situation where the source, the gravitating mass, and the observer are exactly aligned, and the geometry has a symmetry about this axis, we should see *infinitely* many such images all distributed along a ring perpendicular to the axis. This ring is sometimes called the *Einstein ring* (see Figure 5-13).

In the 1960s Jeno Barnothy and S. Refsdal had independently suggested that such illusions might well confront the astronomer looking at remote sources of radiation. One could call them examples of *gravitational lensing*. However, these possibilities were then considered amusing curiosities and no more. But the situation changed dramatically in 1979 when a group of astronomers, D. Walsh, R. F. Carswell and R. J. Weymann, came up with a startling case precisely of this kind. The case concerned a pair of quasars with the catalogue names 0957+561 A & B. To understand the significance of this observation let us first see what quasars are.

Quasars are extremely powerful but very compact sources of light, so much so that before their significance was appreciated in the 1960s, a few bright quasars were mistaken for stars. Closer studies, however, revealed them to be much more powerful than typical stars and also considerably more distant from us than stars in our Milky Way Galaxy.

The quasars 0957+561 A & B were known to be such powerful and distant sources, but with the peculiarity that their directions as seen from here were very close to each other. Only a small angle of 6 arc-seconds separated *A* from *B*. However, when Walsh, Carswell, and Weymann studied them more closely they found considerable similarity in their spectra and other features. While there is nothing to prevent two very similar looking objects from appearing near each other, like human twins, one expects such quasar twins to be rather rare.

This is where the above gravitational lensing idea begins to assert itself. One could ask, Instead of there being two identical looking quasars in reality, are we seeing two images of only one real source? If so, where is the gravitating object *G* of Figure 5-12? Walsh et al. looked and found a galaxy that just about did the trick. Figure 5-14 illustrates the geometry of the entire system. Subsequently, it was noted that this galaxy belongs to a cluster which collectively also contributes to the bending of light.

The discovery demonstrated that gravitational bending of light can very subtly distort the astronomer's perception of reality. After 0957+561 A & B there were other cases of multiple quasars identified as gravitationally lensed images. In some cases three and even four images have been seen. Also, in Figure 5-15, we see an arc extending across several galaxies. Lying in the Abell cluster 370, this arc could be part of the Einstein ring! Another characteristic

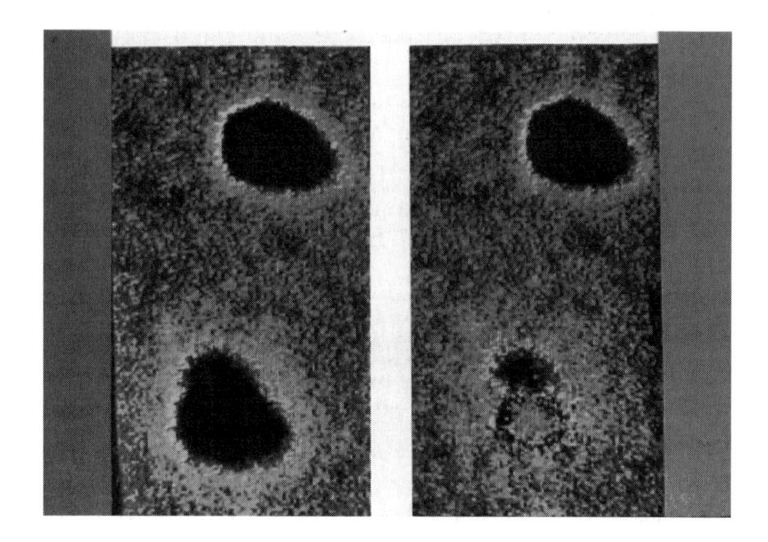

Figure 5-14. The two quasars 0957+561 A & B are shown on the left. The two images are interpreted as arising from a single source, by graviational lensing. The figure on the right shows the lensing galaxy (in the lower part) after subtracting the upper quasar image from the lower one. (Adapted from the photograph by Allan N. Stockton, Institute of Astronomy and Planetary Geosciences Data Processing Facility.)

of lensing is that it may brighten one of the images by a large factor by concentrating a number of light rays together. In Chapter 10 we will see how this effect has been cleverly used by observers.

Is there any way by which an astronomer can check whether two closely situated quasars are distinct real entities or images of one source? If the light from the quasars fluctuates with time then the answer is, yes, possibly. For, even though the images are nearly identical, the lengths of their light paths are in general different. Thus, in Figure 5-12 the path which gave the image $S_1$ may be longer than the path for the other image, in which case we are seeing in $S_1$ an image of the source at an *earlier* epoch than in $S_2$. Thus, the light variations seen in $S_1$ should be repeated in $S_2$ after a specific time gap, like a few weeks or a few months or so. Hence, if we continuously monitor both $S_1$ and $S_2$, we may be able to match their light fluctuation pattern by imposing a suitable time gap.

Attempts like this are being made to discover whether the light

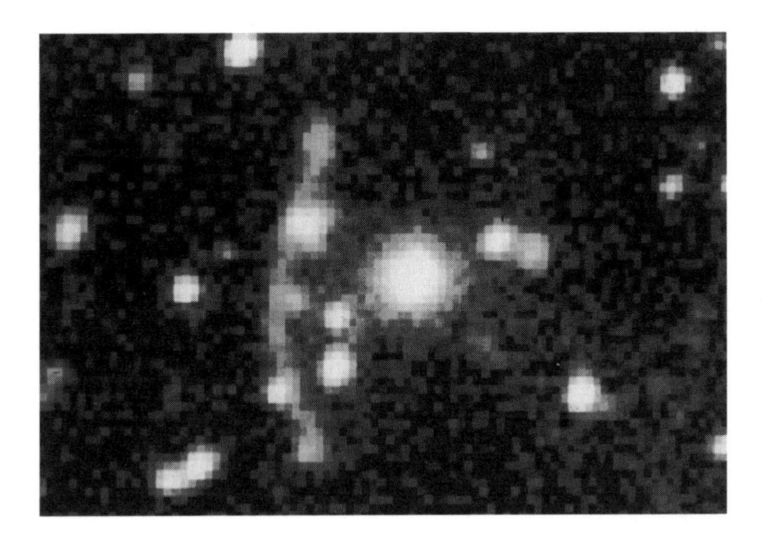

Figure 5-15. An arc seen in the Abell cluster 370 could be an instance of gravitational lensing. (Adapted from figure 1 of the article 'Arc(let)s in clusters of galaxies' by B. Fort and Y. Mellier, *Astronomy and Astrophysics Review* 5, 239, 1994.)

variations of the two images 0957+561 A & B can be so matched. There are some indications that such matching can be done but it is still too early to say that the result is confirmed.

Whatever the outcome of these investigations, the mere possibility of gravitational lensing has warned the astronomer against the old adage 'Seeing is believing'.

# 6
# *Ocean tides and gravity waves*

## WHEN NEWTON AND EINSTEIN AGREE

Einstein's general theory of relativity and Newton's law of gravitation offer radically different interpretations of the phenomenon of gravity. Yet, in practical terms, the difference between their predictions seem to be very small. In Chapter 5 we saw two examples of observations in the solar system: the precession of the orbit of Mercury and the bending of light rays from a distant star by the Sun. In both cases the differences in the predictions of Newton and Einstein are very small and are measurable only with very patient and sophisticated astronomical observations. Is it just a coincidence that these two approaches give almost the same answer?

A mathematical analysis of Einstein's equations tells us that the agreement between the two approaches is not coincidental. It can be shown that, in *all* phenomena of *weak* gravitational effects and where the gravitating bodies are moving slowly compared to light, the two theories must almost agree. In our discussion of the escape speed in Chapter 3, we saw how to measure the relative strength of gravity. We use the criterion of the escape speed in the present context to understand the difference between 'weak' and 'strong' gravity. The rule is simple: compare the escape speed $V$ with the speed of light $c$. If the ratio $V/c$ is very small compared to 1, the gravitational effects are weak. If the ratio is comparable to 1, say between 0.1 and 1, the gravitational effects are strong (see Figure 6-1). Referring back to Table 3-2, we see that the gravitational effects are weak in all cases except on the surface of neutron stars.

Figure 6-1. When the escape speed $V$ is small compared to the speed of light $c$ – that is, when we are dealing with situations of weak gravity – and when objects are moving slowly compared to $c$, Newton and Einstein agree. When $V$ becomes comparable to $c$, the disagreement between the two theories becomes significant.

This is the reason why, in spite of the conceptual and observational superiority of Einstein's theory, Newton's law is still usable. Indeed, because the mathematical formalism of general relativity is much more complicated than the Newtonian framework, astronomers *prefer* to use the latter in cases of weak gravity. In this chapter, we shall describe a few phenomena of gravity and for this we will use the Newtonian framework, except toward the end. In subsequent chapters, however, we shall follow Einstein's framework, for where the difference between the approaches of Newton and Einstein becomes significant we should use what we believe to be the more reliable of the two frameworks.

## THE TIDAL FORCE

Consider first the following situation illustrated in Figure 6-2. $M$ is a gravitating spherical mass exercising its force of attraction on three equal point masses $A, B$, and $C$. As shown in Figure 6-2a, the masses $A, B$, and $C$ are located in a straight line with $C$ in the

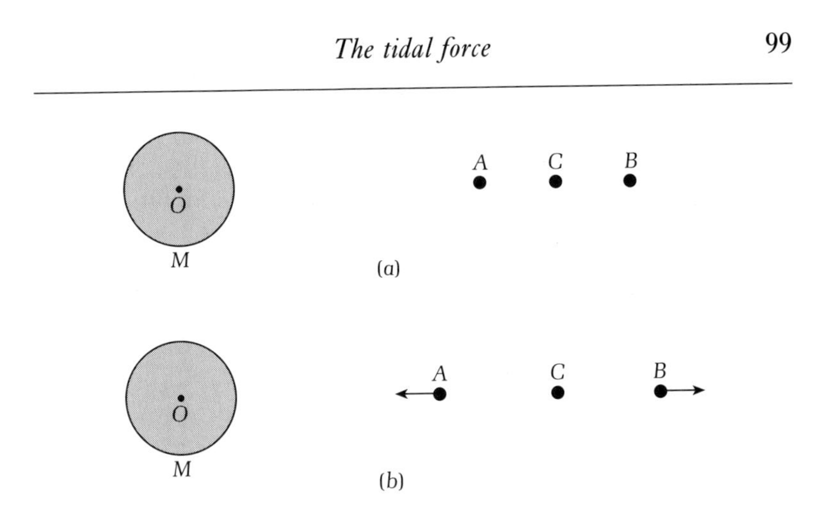


Figure 6-2. When a force of just the right magnitude is applied to hold $C$ where it is, the same force is inadequate to hold $A$ and more than adequate to hold $B$ where they are. In either case, $A$ and $B$ move away from $C$.

middle. To illustrate the point we want to make, we take the simple case of the line $ACB$ passing through the centre $O$ of $M$.

From Newton's law, we know that $M$ attracts all three of the masses $A, B, C$ as if all its mass were concentrated at $O$. We also know that, by the inverse-square law, the acceleration produced by this force of attraction is the largest for $A$ (which is closest to $O$) and the least for $B$ (which is farthest from $O$). The tendency of this force is of course to make $A, B$, and $C$ all fall toward $O$.

Now suppose that we wish to check this tendency by applying a force in the opposite direction. Let us apply an equal and opposite force on $C$ that just holds it at rest. If we apply the *same* force on $A$, it will not be sufficient to counter the force of gravity from $M$. As a result $A$ *will still move toward* $O$. However, if we apply the same force on $B$, it will exceed the force of gravity on $B$, as a result of which $B$ *will move away from* $O$. In other words, the masses $A$ and $B$ both move away from $C$. The length $AB$ tends to increase, as shown in Figure 6-2b.

If the particles $A, B$, and $C$ were connected and were part of the same body, the effect of our example would be to stretch the length $AB$. This stretching force is known as the *tidal force* exerted by $M$ on the body. Besides exerting a force of attraction, $M$ also tends to *distort* the shape of the body through this tidal force.

The name *tidal force* has come from its application to the tides on the Earth. In the example just discussed, let $M$ stand for the Moon and let the system of three particles be replaced by the Earth. We then encounter the situation shown in Figure 6-3.

We see in Figure 6-3 that the dashed line shows the tendency of the Earth to become elongated along the line toward (and away from) the centre of the Moon. The rigidity of the Earth's solid crust is strong enough to prevent a bulging in response to the tidal force. However, where the surface happens to be covered with water it is a different matter. Unlike the rigid crust, the water can and does move. Thus, the effects of this force show in the motions of the ocean tides. The tides are strongest at the side facing the Moon.

The effect is naturally magnified if the Sun also happens to lie approximately on the Earth–Moon line. This happens at Full Moon and New Moon. On these occasions, the tides in the ocean are the most spectacular.

Where does the energy of motion of the water come from in an ocean tide? Obviously, the source of energy lies in the tidal force of gravity exerted by the Moon on the Earth. In these days of energy shortages, we cannot afford to ignore natural sources of energy, and the ocean tides represent one such natural source.

## THE EXPERIMENT OF TWEEDLEDUM AND TWEEDLEDEE

In fact, the tidal force is a mechanism whereby energy can be transferred from one astronomical object to another. An interesting thought experiment illustrating this process was once described by two astrophysicists, Hermann Bondi and William McCrea. The experiment concerns two remarkable creatures called Tweedledum and Tweedledee, so named after the famous characters in the Lewis Carroll novel. They are made of pliable material that allows them to change shape. Originally, they had identical spherical shapes and were set moving around each other in highly eccentric orbits under their mutual gravitational force. They were given an order that they must always move along these orbits. Let us assume further that both Tweedledum and Tweedledee always have shapes symmetrical about axes perpendicular to the plane of their orbits.

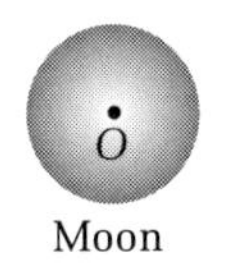


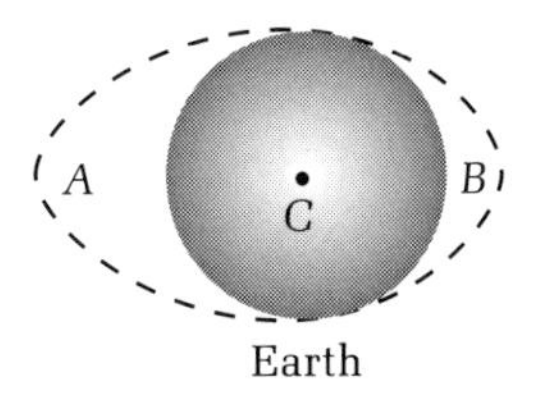


Figure 6-3. The dashed lines indicate the Earth's tendency to bulge at ends $A$ and $B$ along the line joining the centres of the Earth and the Moon. Compare this situation with that described in Figure 6-2. In both cases the tidal effect is larger at $A$ than at $B$.

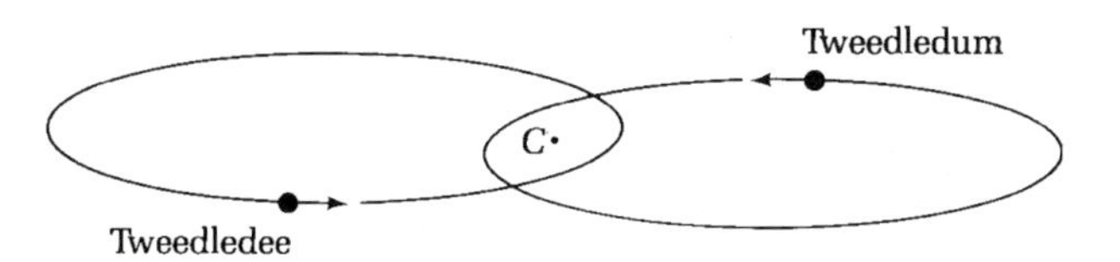


Figure 6-4. The highly eccentric elliptical orbits of Tweedledum and Tweedledee. They move in such a way that their centre of mass, which is the point $C$ midway between them, remains fixed in space.

Figure 6-4 illustrates the orbits of Tweedledum and Tweedledee. The orbits are identical, and the two move along their respective orbits in such a way that the point $C$ midway between their centres is fixed. This is characteristic of the way any two objects move in each other's gravitational attraction. Even in the case of the Earth and the Sun, although we usually talk of the motion of the Earth only, the Sun should in principle also move under the Earth's gravitational force. However, in the Earth–Sun system, because the Sun's mass greatly exceeds that of the Earth, the Sun's motion is negligible. The centre of mass of the Earth and the Sun in fact lies within the Sun.

But to return to Tweedledum and Tweedledee: they would have no difficulty obeying the command to move along their assigned orbits if they were rigid creatures. Since they are pliable, their shapes get distorted by each other's tidal forces. Let us investigate this effect further.

In Figure 6-5a, we have two spheres $A$ and $B$. We saw earlier that, if $B$ is subject to $A$'s tidal force, it will bulge in the direction

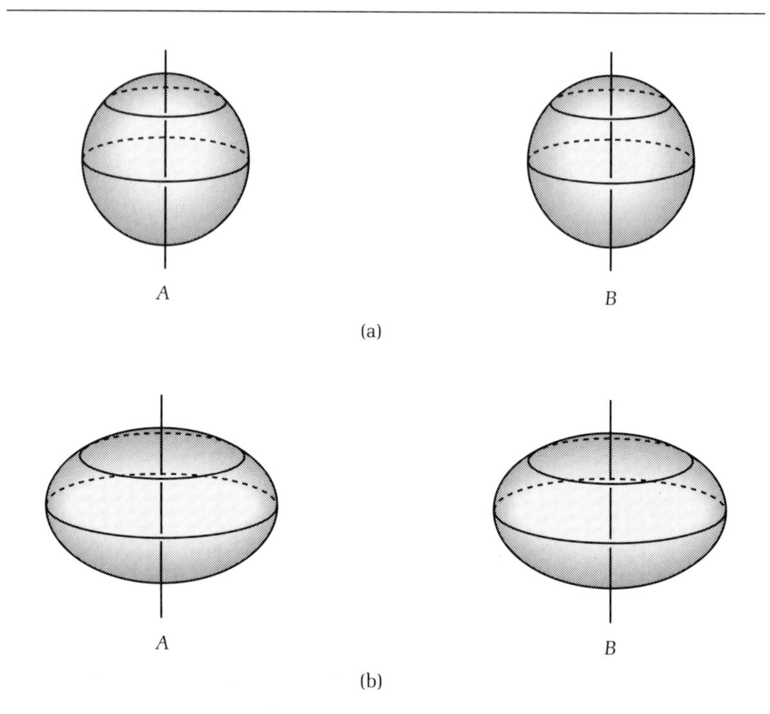


Figure 6-5. In (a) we have two spherical objects *A* and *B*. Because of their gravitational tidal force on each other, both tend to bulge in the middle. If they are constrained to remain symmetric about an axis as in (b), they become oblate spheroids.

of *A*. If *B* is also constrained to be symmetrical about an axis perpendicular to this direction (as Tweedledum and Tweedledee are), it will bulge all along its equator, as shown in Figure 6-5b. The same happens to *A* under the tidal force of *B*.

When a sphere bulges along its equator as in Figure 6-5b, it becomes an *oblate spheroid.* Had *B* instead been elongated at the poles, it would have become a *prolate spheroid*, as shown in Figure 6-6. An oblate spheroid is bun-shaped, and a prolate spheroid is egg-shaped, more like a rugby ball.

In the situation described in Figure 6-5b, when the two spheres become spheroids their mutual gravitational attraction *increases.* This is because *A* attracts *B* more powerfully as an oblate spheroid than as a sphere. This is the reason why the job of Tweedledum

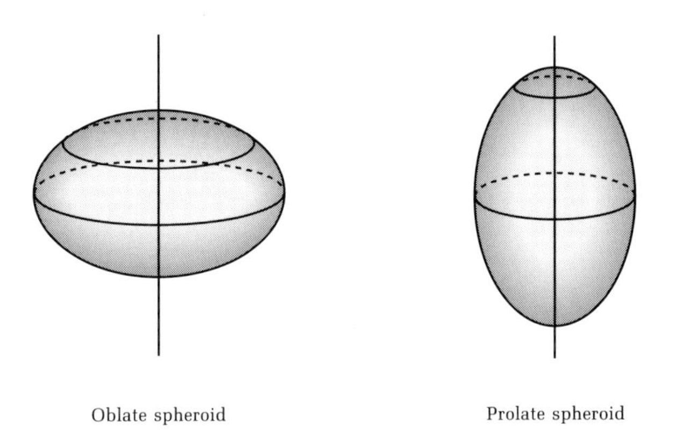


Figure 6-6. An oblate and a prolate spheroid.

and Tweedledee is made more difficult – as they go around each other, their shapes tend to become oblate and the force between them increases, an effect which in turn tends to distort their orbits. To overcome this difficulty, one course is open to them. If one of them becomes oblate, the other must become prolate, because the prolate spheroid exerts less force on its companion than a sphere.

Now Tweedledum is clever and unscrupulous whereas Tweedledee is stupid and simple. Tweedledum uses his intelligence to capitalize on the situation in the following way. He makes Tweedledee sign an unfair agreement saying that whenever Tweedledum becomes oblate, Tweedledee must become prolate, and vice versa. The exact shape that Tweedledee must acquire will be calculated and communicated to him by Tweedledum. The basis of the calculation is the criterion that their orbits must not change. Not knowing how to make this calculation himself, Tweedledee simply signs along the dotted line. This is a mistake for which he will have to pay dearly, as we will now see.

Moving in highly eccentric orbits, Tweedledum and Tweedledee alternately come close together and move farther apart. In Figure 6-7a, they are shown near each other. At this stage, there are huge tidal forces between them, and here Tweedledum allows himself to become oblate. To achieve this, he of course does not have to do any work; as shown in Figure 6-5, the tidal force does the work of

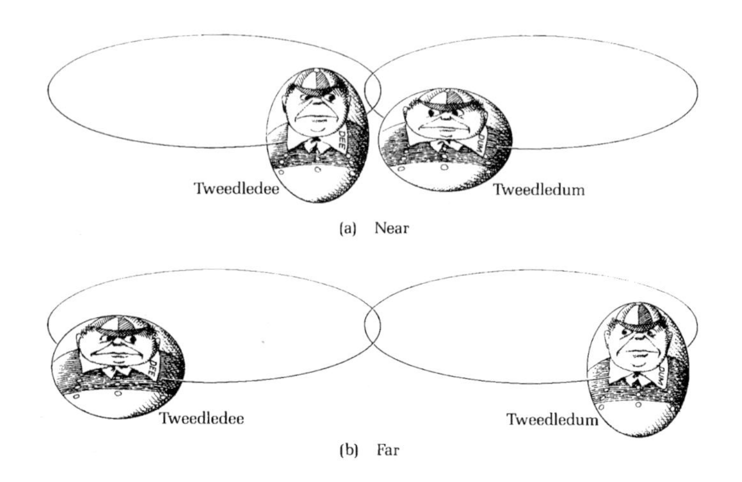


Figure 6-7. Tweedledum gains energy at the expense of Tweedledee by changing shape in accordance with tidal forces.

changing his shape. *And this work done by an outside force increases Tweedledum's energy reservoir.* What about Tweedledee? He must go into an appropriate prolate shape as per the agreement he has signed. To do this, he has to work against the tidal force, and as a consequence his energy reservoir is depleted.

Meanwhile, the two have started moving apart. When their separation is close to the maximum, as in Figure 6-7b, Tweedledum changes himself to a prolate shape, thereby making Tweedledee go oblate. However, when the two are far apart, the tidal force between them is weak. Hence, even though Tweedledum does some work to become prolate, he has to spend very little energy doing so. Correspondingly, although Tweedledee gains some energy because this time the tidal forces *help* him change his shape, the energy gained by him is small.

We now see how unfair this agreement is. During each orbit, Tweedledum alternately gains a lot of energy and spends very little, while Tweedledee loses a lot of energy and gains little. What is happening is that Tweedledum is cleverly using the tidal force to extort energy from poor Tweedledee, although the latter cannot, on the face of it, see where the unfairness of the agreement lies. The reader may be clever enough, however, to see a parallel between this scenario and occasional unfair trade agreements between countries!

## TIDAL DISRUPTION

The crucial property of the tidal force that leads to the remarkable effects of the Tweedledum–Tweedledee experiment is that its strength diminishes with distance very rapidly. In fact, its strength diminishes in inverse proportion to the *cube* of the distance. Thus, if the farthest distance of separation between Tweedledum and Tweedledee is ten times the distance of their closest separation, the tidal force in the former case is one-thousandth of what it would be in the latter case. This is why Tweedledee had to work so much harder than Tweedledum in changing from oblate to prolate form.

We therefore expect the tidal force to be very powerful in circumstances where two astronomical objects are close to each other. How close can a planet be to its parent star? How close can a satellite come to the planet around which it revolves? The answers to these questions invariably involve calculations of the tidal force. If a satellite comes too close to a planet, the tidal force may become so enormous that it destroys the satellite. The same consideration applies to a planet near its parent star. The limit beyond which the tidal forces become highly disruptive is known as the *Roche limit*, so named after the work of E. Roche (1820–1883).

Another circumstance in which the tidal force becomes significant is that of a binary star system. Like Tweedledum and Tweedledee, two stars in a binary system go around each other in elliptical orbits. A typical binary system is shown in Figure 6-8a. The stars $A$ and $B$ (which need not be and usually are not of identical mass) orbit around their common centre of mass. The dashed line shows the so-called *Roche lobe*. If either star becomes large enough to cross its Roche lobe, the tidal force of its companion will begin to make its disruptive presence felt.

We saw briefly in Chapter 4 that, as a star evolves, it passes through the red-giant phase and becomes very large. If one of the stars in our binary system becomes a red giant, its outer surface begins to spill over its Roche lobe, and the situation shown in Figure 6-8b will occur. The companion star $B$ now starts exerting a tidal

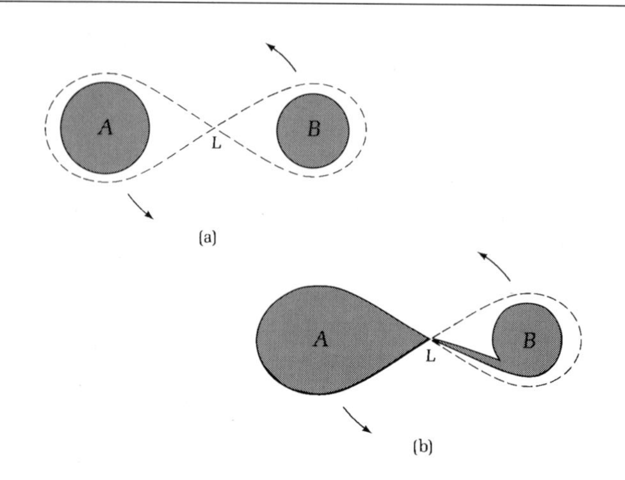


Figure 6-8. The dashed curve describes the Roche lobe of the double star system. The point $L$ is called a Lagrangian point. In (a), stars $A$ and $B$ are well inside their Roche lobe. In (b), star $A$ has expanded and filled its Roche lobe. The tidal forces of $B$ now pull out material from $A$, which escapes through $L$ and falls into $B$. The arrows indicate the direction in which the binary system is rotating.

force on $A$ in such a way as to pull material from the surface of $A$ that faces it. This surface, therefore, becomes distorted, and the material 'falls' toward $B$. We will return to this interesting situation in Chapter 8, for it has potentially dramatic consequences.

## GRAVITATIONAL RADIATION

The effects we have been discussing so far could be analysed within the framework of general relativity in much the same way that they are analysed within the Newtonian framework. We have used the Newtonian framework to describe these effects since it is intuitively the simpler of the two to grasp. Now, however, we turn to another effect of weak gravity that can only be described within the relativistic framework. This is the phenomenon of gravitational radiation. It arises when the motions, even in a weak gravity environment, are significant enough to vitiate the Newtonian framework.

Recall from Chapter 5 the criticism of Newton's law of gravitation, that gravitational effects supposedly propagate instantaneously, that is, with *infinite speed*. According to Einstein's special theory of relativity, no physical effects can be communicated across space faster than the speed of light. Does Einstein's theory of gravitation as stated in the general theory of relativity meet this requirement?

The answer to this question is 'yes', but the details of how the gravitational effects propagate from $A$ to $B$ are immensely complicated when we are dealing with the situations where these effects are strong. However, for weak gravitational effects, the situation is considerably simpler and is analogous to the more familiar case of electromagnetic radiation. Like electromagnetic radiation, one can talk of *gravitational radiation.*

The simplest system generating electromagnetic radiation is one where an electric charge oscillates, or moves to and fro much like the bob of a simple pendulum. As shown in Figure 6-9, the radiation emanating from such a motion of the electric charge consists of electrical and magnetic disturbances that propagate away from the charge *with the speed of light.* The disturbances generated on the initially calm surface of a pond of water when a peeble is dropped into it are roughly analogous to electromagnetic radiation.

The electrical and magnetic disturbances carry energy. Where does this energy come from? Its sources can be traced to the motion of the charge. As it radiates electromagnetic waves, the electric charge tends to lose its kinetic energy and hence experiences a damping of its motion. Just as the bob of the pendulum eventually slows to a state of rest because of the damping effect of air resistance, so our oscillating charge also slows down because of the damping produced by radiation.

The gravitational analogue of the situation is described in Figure 6-10. This is the now-familiar binary star system. As the stars go around each other, they generate disturbances in the geometry of spacetime. These disturbances propagate out to large distances *with the speed of light.* And as in the case of the electric charge, these disturbances carry energy, which causes a damping of the binary star motion. Because of this damping, the stars come closer and closer (their orbits shrink), and the angular speed with which they go around each other increases.

Has gravitational radiation been detected in a laboratory experiment? Here we begin to depart from our analogy with electromagnetic radiation. Unlike the electromagnetic case, gravitational

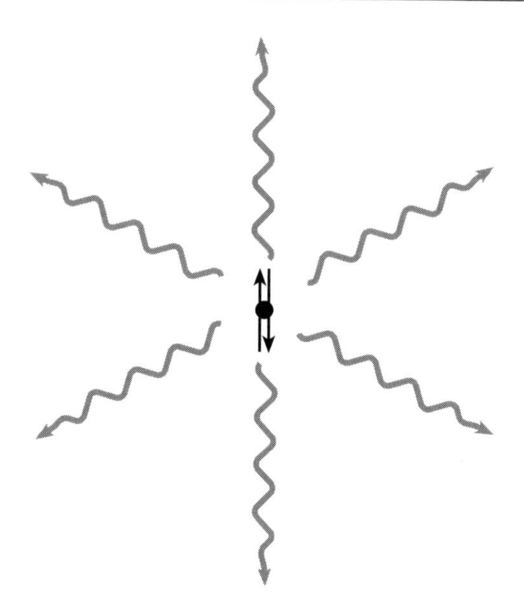

Figure 6-9. A schematic representation of radiation from the to and fro motion of an electric charge. Such a motion generates disturbances of electricity and magnetism around the charge. These disturbances move as waves in directions radially away from the moving charge.

radiation is so feeble that production of gravity waves lies beyond the capabilities of the terrestrial laboratory. To get some idea of the problem, imagine a cylindrical beam of radius 1 metre and length 20 metres, density 7.8 times that of water, and a mass of about 490 tons (see Figure 6-11). To produce gravitational radiation, we rotate this beam about its middle with an angular velocity of just under 4.5 revolutions per second. (For a rate of revolution faster than this the rod will not be able to retain its shape under the tensile strains generated.) To generate gravitational radiation of 1 watt of power, we need about 1 million million million million million such cylindrical sources! For realistic sources of gravitational radiation one must therefore look to distant astronomical objects.

There is indeed some indirect evidence for the existence of gravitational radiation from astronomical sources. In early 1979, J. H. Taylor, L. A. Fowler, and P. M. McCulloch, observing from the 1000-foot dish at the radio astronomy observatory in Arecibo, Puerto

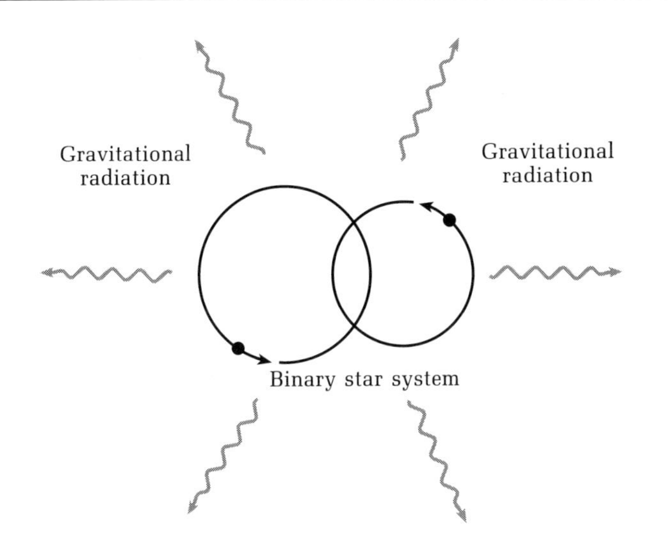


Figure 6-10. A schematic representation of gravitational radiation by a binary star system.

Rico, reported that their monitoring of the motion of the binary pulsar PSR 1913+16, consisting of the pulsar and another compact star, showed the effects expected from gravitational radiation. (The letters PSR indicate that the source is a pulsar; the numbers following these letters fix the direction of the pulsar in the sky.) In particular, the orbital period of the two members around each other *decreases* at just the rate expected on the basis of gravitational radiation. The crucial point is that the theory of relativity gives the right answer for the expected rate of decrease of orbital period and performs better than a group of competing theories of gravity. Hence, the observation has boosted confidence in relativity.

The binary pulsar PSR 1913+16 was discovered by Russell Hulse and Joe Taylor back in 1978.* It not only has the advantage that, being a binary system, it provides a means of checking gravity in operation, but it also has the pulsar as a very precise timekeeper. A pulsar is believed to be a highly dense star made mostly of neutrons that is left over as the surviving core of a supernova explosion (see the discussion on neutron stars in the next chapter). The star spins

* For this discovery Hulse and Taylor shared a Nobel Prize in 1993.

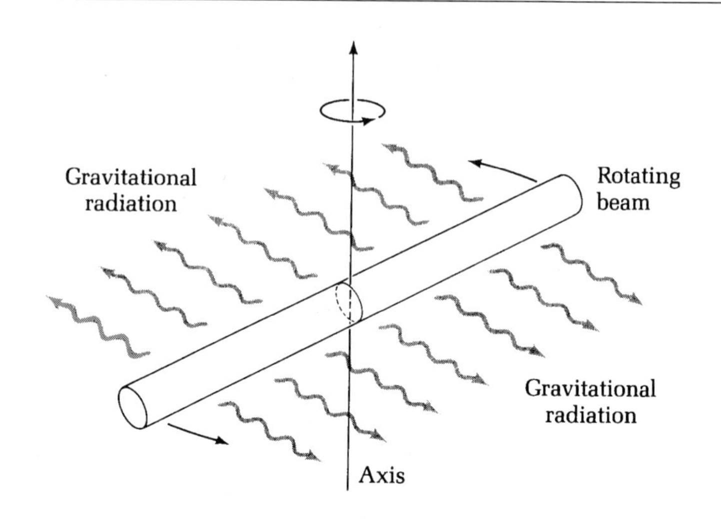


Figure 6-11. A laboratory radiator of gravitational waves. The cylindrical beam rotating about an axis perpendicular to its length will radiate gravity waves. The effect is, however, very, very small.

rapidly and in the process emits electromagnetic pulses with extreme regularity. The Hulse–Taylor pulsar emits pulses with a period of 0.059 029 995 270 9 second. The accuracy to thirteen decimal places underscores the regularity of pulsars as timekeepers.

Because of a unique combination of the above two features, a binary pulsar like PSR 1913+16 is an admirable cosmic laboratory for testing theories of gravity. We have already mentioned the test involving gravity waves. There are others. For example, just as in the case of the precession of the perihelion of the planet Mercury (see Chapter 5), here we see a similar effect. The direction along which the two orbiting stars come closest to each other is found to change with time. (The Newtonian law of gravitation predicts this direction to be stationary.) The observed rate of change of this direction is considerably faster than in the case of Mercury; it is about 4.2° per year. If it becomes possible to estimate the masses of the two stars accurately, we can then compare this observed precession rate with a theoretically predicted rate. Although we have approximate answers on both the observational and theoretical fronts the agreement between the two looks good.

In any case, PSR 1913+16 is another example of how accurate astronomical measurements made possible by our present technology can detect even the weak effects of gravity.

## GRAVITY WAVE DETECTORS

The Hulse–Taylor pulsar underscores the importance of detecting gravitational radiation directly. The observations of a decreasing binary period tell us indirectly that the system is emitting gravitational radiation. However, we have as yet no *direct* means of detecting that radiation. So far as the astronomy of gravity waves is concerned, we are essentially in the same state that the astronomers were in during the 1950s regarding X-ray astronomy. The astronomers felt that there should be cosmic sources of X-rays, yet they had no telescopes or detectors to detect them. For, because of absorption by the Earth's atmosphere, no such radiation reaches a ground based detector. Only after space technology came into its own could such detectors be built and launched in satellites above the atmosphere.

Likewise, for gravitational radiation we need highly sensitive detectors that can pick up even a very weak signal like that from the Hulse–Taylor pulsar. How does such a detector function?

Recall that according to general relativity the effects of gravity manifest themselves through spacetime geometry (Chapter 5). As a gravity wave passes through space, it too brings about tiny changes in the geometry of the region. And these changes manifest themselves through changes in spatial and temporal measurements.

Joe Weber, who pioneered the detector technology in the 1960s, sought to measure these changes as a gravity wave passes through a large metallic bar. Any two points in the bar would 'notice' this effect through the change in their spatial distance for example. By an ingenious device these changes can be converted into electrical signals which are then measured.

The Weber 'bar antenna', however, turns out to have limits to its sensitivity which makes it a not very efficient detector in the energy range of gravity waves expected from typical cosmic sources, such as compact binaries. A new technology using laser interferometry offered better sensitivities and became more popular with the gravity wave experimentalists in the 1980s.

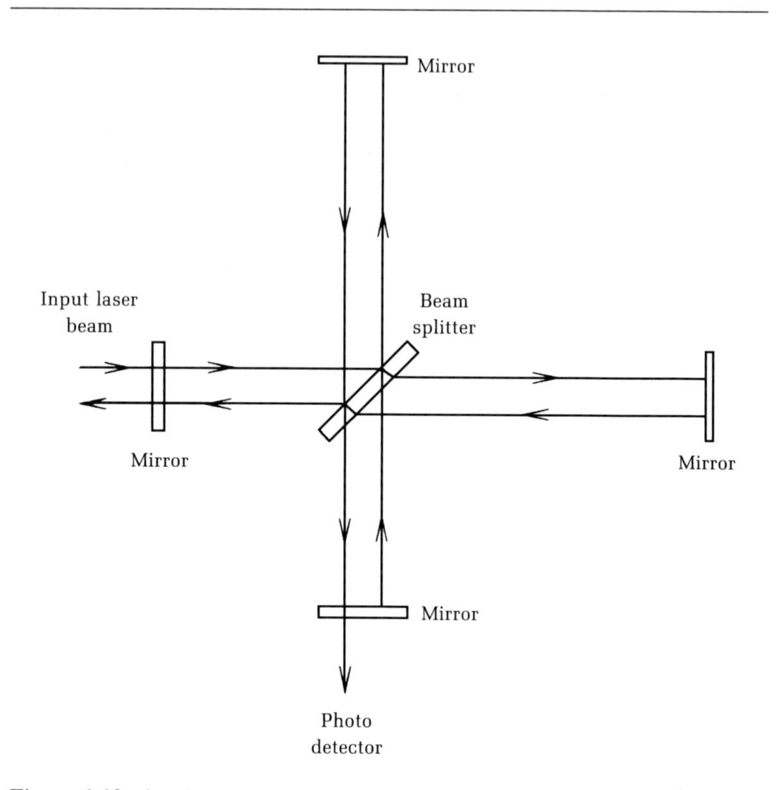


Figure 6-12. A schematic representation of a laser interferometer detector of gravity waves.

A laser interferometer for gravity waves works on a principle similar to that of an optical interferometer like that shown in Figure 5-4. In Figure 6-12 we see a schematic drawing of a laser interferometer detector of gravity waves, which, ideally, can spread over several kilometres in either direction. When a gravity wave crosses its arms, their dimensions change slightly, leading to interference of laser beams which can be measured. The sensitivity demanded by the problem places stringent limits on the specifications of the various components of the interferometer. Thus the laser beam must be 'pinpointed' and should travel in tunnels with a very high degree of vacuum, and the resulting data must be subjected to very sophisticated mathematical analysis to extract the faint signal from a jumble of noise.

These demands of accuracy have not daunted scientists from many countries in their enterprises of creating accurate, stable lasers, attaining high vacuum over long distances, and evolving algorithms that will most effectively search for the proverbial needle in a haystack.

It is still an open bet whether gravity waves from astronomical sources will be detected here on the Earth during the twentieth century!

# 7

# *The strange world of black holes*

## THE BLACK HOLE IN HISTORY AND ASTRONOMY

The oldest mention of a black hole is found not in books of physics or astronomy but in books of history. In the summer of the year 1757, Nawab Siraj-Uddaula, the ruler of Bengal in eastern India, marched on Calcutta to settle a feud with the British East India Company. The small garrison stationed in Fort William at Calcutta was hardly a match for the Nawab's army of 50 000. In the four-day battle that ensued, the East India Company lost many lives, and a good many, including the company's governor, simply deserted. The survivors had to face the macabre incident now known as the *Black Hole of Calcutta*.

The infuriated Nawab, whose army had lost thousands of lives in the battle, ordered the survivors to be imprisoned in what came to be known as the Black Hole, a prison cell in Fort William. In a room 18 feet by 14 feet, normally used for housing three or four drunken soldiers, the 146 unfortunate survivors were imprisoned. The room had only two small windows (see Figure 7-1). During the 10 hours of imprisonment, from 8 p.m. on 20 June to 6 a.m. on 21 June in the hottest part of the year, 123 prisoners died. Only 22 men and 1 woman lived to tell the tale.

Apart from its macabre aspect, the Black Hole of Calcutta did bear some similarity to its astronomical counterpart, involving as it did a large concentration of matter in a small space from which there was no escape. While discussing the notion of escape speed in Chapter 3, we made a passing reference to the concept of a black

Figure 7-1. The Black Hole of Calcutta. (From *The Black Hole of Calcutta*, by Noel Barber, London: Collins.)

hole. We now define it as an object whose mass concentration generates a gravitational attraction so strong that not even light can escape from its surface. Who first thought of such a bizarre object?

The French mathematician Laplace, who had carried out exhaustive research on Newtonian gravity (see Chapter 2), conceived of the notion of such a black hole in the year 1799. Although Laplace did not use the term 'black hole' to describe these objects, it is clear from his discussion that his concept implied the property of the escape speed exceeding the speed of light. For a spherical object of mass $M$ and radius $R$, this property implies that $R$ cannot exceed the value

$$R_S = \frac{2GM}{c^2}$$

where $c$ is the speed of light and $R_S$ is the critical value for the radius of a Laplacian black hole. (The meaning of the subscript S will become clear when we discuss black holes within the framework of general relativity.)

To get an idea of how small $R_S$ is, let us look at a couple of examples. Suppose we shrink the Earth until it satisfies the above criterion for a black hole. In this case, $R_S$ is of the order of 8 millimetres! That is, the radius of the Earth must be less than this value if it is to become a black hole. If we consider a somewhat more massive object, the Sun, the corresponding value of $R_S$ is about 3 kilometres. The actual radius of the Sun is nearly quarter of a million times $R_S$.

However, Laplace was not the first physicist to talk about black holes in this way. An English physicist named John Michell published a paper in 1784 in the *Philosophical Transactions of the Royal Society of London* (vol. 84, p. 35), with the title 'On the Means of Discovering the Distance, Magnitude, etc. of the Fixed Stars, in Consequence of the Diminution of the Velocity of the Light, in case of such a Diminution should be found to take place in any one of them, and such other Data should be procured from Observations, as would be further Necessary for that Purpose.' In this paper Michell discussed stars in general and speculated about the possibility of astronomical objects whose mass and radius satisfy the criteria for a black hole.

These speculations, and the work of Laplace, were based on the Newtonian theory of gravitation. We have already seen, when discussing the effects of strong gravity, that the Newtonian framework is suspect and that modern physicists prefer to use Einstein's general theory of relativity. Let us therefore discuss the phenomenon of black holes within Einstein's framework. We will find that black holes are even more dramatic within this framework than within the Newtonian framework.

## HOW ARE BLACK HOLES FORMED?

It is already clear from the two examples of the Earth and the Sun that even within the Newtonian framework black holes are highly unusual objects. Even astronomers accustomed to dealing with extreme states of matter find black holes esoteric.

As we saw in Table 3-2, the stronger the gravitational attraction of an object, the higher is the velocity required to escape from it. For stars and planets, the escape speed is a small fraction of the speed of light. Even for neutron stars, the escape speed does not exceed two-thirds the speed of light.

Neutron stars, however, suggest a clue to the formation of black holes. In a neutron star, the density of matter reaches a value as high as a *million billion times* the density of water. What force is responsible for holding a neutron star in equilibrium? We have seen in Chapter 4 how powerful the force of gravity is inside a star. The Sun, for example, would not be able to maintain its present size against the contracting tendency of its own gravitational force if it were not for the thermal pressures within it. The thermonuclear reactions that take place in the deep interior of the Sun not only generate enough energy to keep the Sun shining but also provide a stabilizing outward pressure that keeps the Sun from collapsing. In a neutron star, no thermonuclear reactions take place. In fact, the neutron star is formed toward the *end* of a star's normal lifetime, when it explodes as a supernova.

In Chapter 4 we carried the story of stellar evolution to this supernova stage. We did not ask what is left behind when the star explodes as a supernova since the question is of greater relevance in the present context than in our discussion in Chapter 4 of the star's role as a fusion reactor.

Our present understanding of the supernova explosion process is by no means perfect, but theoretical work in this area of astrophysics suggests that the central hot core of the star is left as a remnant (see Figure 7-2). This core contracts and becomes a neutron star *if its internal pressures can support the crushing force of the star's own gravity*.

What are the internal pressures in a neutron star? At the high density of up to a million billion times the density of water the matter in the star exists predominantly in the form of neutrons. And these neutrons are closely packed to more than $10^{44}$ neutrons per cubic metre.

The laws of quantum physics describe the behaviour of matter at the very small scale of atoms and nuclei. These laws describe certain restrictions on the way a set of identical particles can be packed in a given volume. When these restrictions (first discovered by the atomic physicist Wolfgang Pauli) are calculated for neutrons packed inside a neutron star, we find that this matter is endowed

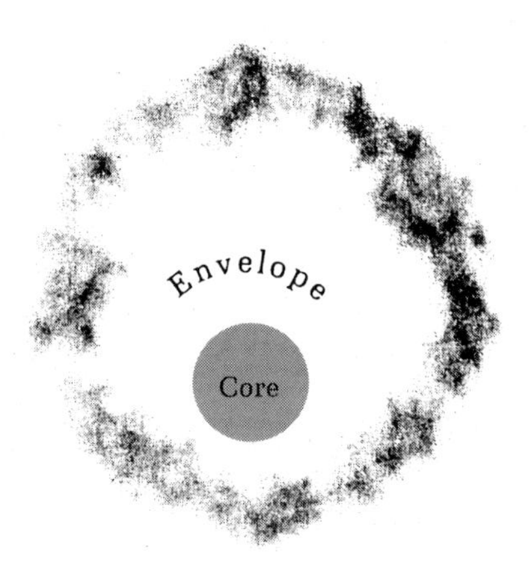


Figure 7-2. The remnant of a supernova explosion may be a neutron star.

with a new kind of pressure. Known as degeneracy pressure it has a tendency to *resist* the closer packing of neutrons and therefore helps maintain the star's equilibrium against its self-gravity.

Does it seem that we have digressed somewhat from our question of how black holes are formed? We appear to have arrived instead at a description of how a neutron star is formed and maintained in equilibrium. No, we have not disgressed from the topic of black holes! We are in fact very close to the scenario that leads to the formation of a black hole.

In our discussion of the internal pressure of a neutron star, we said that degeneracy pressure *helps* to maintain equilibrium. There are limitations on how much weight this type of pressure can be called upon to support. Calculations show, for example, that if the mass of the neutron star exceeds three times the mass of the Sun ($3M_{\odot}$), this pressure cannot be invoked to maintain equilibrium. The mass limit of $3M_{\odot}$ may in fact be an overestimate – it could be as low as $2M_{\odot}$.

What happens if the remnant of a supernova explosion has a mass exceeding this limit? The remnant obviously cannot survive as a neutron star. Its pressures will be inadequate to withstand the crushing force of its own gravity. The star continues to contract and becomes a black hole.

## GRAVITATIONAL COLLAPSE

The idea of degeneracy pressure arising within a closely packed system of particles was known by the mid-1920s. However, the full implications of this pressure within very massive objects like stars were first appreciated by a young Indian astrophysicist, Subrahmanyan Chandrasekhar, in the early 1930s. Chandrasekhar applied his ideas to stars known as *white dwarfs*. Visually very faint and compact in size, a typical white dwarf may have matter a million times as dense as water.

It was earlier believed that all stars, when they run out of their fuel, would end up as white dwarfs. Chandrasekhar, however, found that there is a limit to the mass that can be held in equilibrium through the degeneracy pressure arising from the close packing of electrons. This limit, known as the *Chandrasekhar limit*, is about 1.4 solar masses. If a star is more massive than this limit, its interior does not become degenerate, and the star will begin to shrink under its force of gravity.

Ironically, it was Eddington, the pioneer in the field of stellar structure, who found this result unpalatable. He rejected Chandrasekhar's concept of degeneracy as unphysical, largely on the grounds that it led to the ridiculous situation of an ever-shrinking star. At a meeting of the Royal Astronomical Society in January 1935, where Chandrasekhar presented his results on white dwarfs, Eddington had this to say: '...The star has to go on radiating and radiating and contracting and contracting until, I suppose, it gets to a few km. radius, when gravity becomes strong enough to hold in the radiation, and the star can at last find peace... I think there should be a law of Nature to prevent a star from behaving in this absurd way!...'.

Had Eddington taken Chandrasekhar's results seriously, he would thus have been led to deduce an important result, viz., the formation of a black hole by gravitational contraction. It was only in the mid-1960s that the idea really attained respectability.

Let us examine the process of the contraction of a star to a black hole in some detail. In Figure 7-3, we see a contracting star in two stages. In Stage I, the star has just begun to contract because its internal pressures have proved inadequate to balance the inward

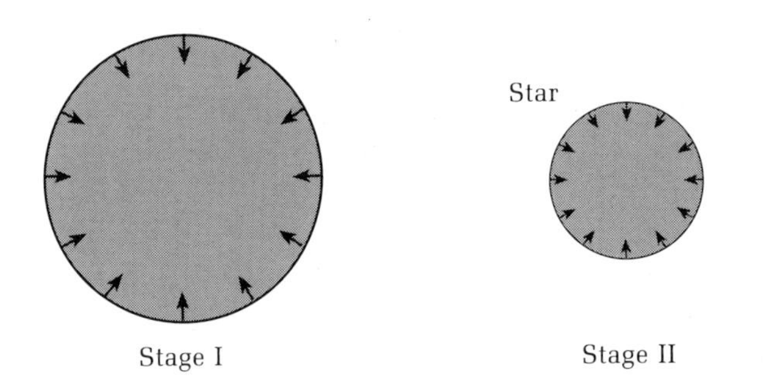


Figure 7-3. In Stage I, gravity is beginning to assert its supremacy over the internal pressures in the star. In Stage II, gravity has far surpassed the internal pressures and is dominating the dynamical behaviour of the star. The star now undergoes gravitational collapse.

force of gravity. The rate of contraction here is slow. In Stage II, at a later time, the star is considerably smaller. According to the Newtonian framework, the force of gravity within the star will grow considerably stronger as it shrinks from Stage I to Stage II, because all its matter has come closer together. Although the star's outward pressure might have grown during this contraction, the outward pressure increases more slowly than the star's inward force of self-gravity. In other words, the imbalance between the two opposing forces in Stage I has increased in Stage II. The star therefore contracts even faster than before, *and this tendency to contract will increase as the star contracts further.*

This situation is often described by the term *gravitational collapse.* From the initial gentle rate of contraction, the star progresses to a catastrophic implosion when its self-gravity becomes so strong that nothing can prevent the collapse of the star.

Of course, we used the Newtonian framework of gravity to explain what is meant by gravitational collapse. Since we are dealing with very strong gravitational effects, we should in fact have used general relativity. Had we done so, in qualitative terms the behaviour of the star would have turned out to be no different. Relativity, however, introduces certain novel features into this situation that we can no longer ignore.

## GRAVITATIONAL REDSHIFT

Let us recall that our discussion of the non-Euclidean geometry of spacetime in Chapter 5 was limited to measurements in space only. This is an opportune time to describe the effects of non-Euclidean geometry on time measurements.

Figure 7-4a shows a spacetime diagram for the imaginary situation where *no gravity* exists – that is, where spacetime is flat. The lines $a$ and $b$ are the world lines of two observers at rest, $A$ and $B$. Suppose $B$ sends light signals to $A$ every second as measured by his own clock. The dashed lines describe the tracks of light rays leaving $B$ and reaching $A$. At what intervals does $A$ receive these signals? These light tracks and the world lines $a$ and $b$ form a succession of parallelograms. In any parallelogram, the opposite sides are equal, and so as measured by $A$'s clock the signals from $B$ will arrive at one-second intervals.

This conclusion is based on Euclid's geometry and does not apply to the situation illustrated in Figure 7-4b, which shows curved spacetime around a massive spherical object. The geometry of curved spacetime was first determined by Karl Schwarzschild (see Chapter 5). The shaded region indicates the presence of the massive object. The signal emitter $B$ is now on the boundary of this object, while $A$ is far away. As before, $A$ and $B$ are at rest relative to the massive object, and the intervals along their respective world lines $a$ and $b$ denote the times measured by their respective clocks. The dashed lines denote the light tracks from $B$ to $A$.

On the plane of the paper, these tracks do not appear straight. However, the rules of geometry are no longer those of Euclid. By the rules of Schwarzschild's geometry, these dashed lines are *straight* light tracks. However, their intercepts on $A$'s world line are longer than those on $B$'s world line. The difference in length means that, although $B$ sends signals at one-second intervals by his clock, $A$ has to wait longer than one second, say $(1 + z)$ seconds, between successive signals from $B$. This fractional increase of $z$ is known as the *redshift* for reasons which will shortly become clear.

Suppose that, instead of sending signals to $A$ every second, $B$ sends light waves of a specific frequency $\nu$ all the time. This means that in every one-second interval, $\nu$ wave crests will leave $B$ toward

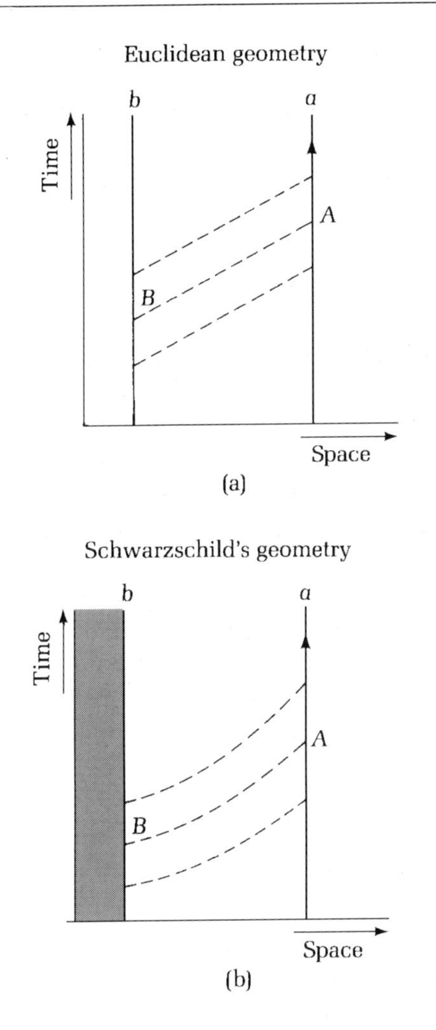


Figure 7-4. Two examples of spacetime geometry: (a) Euclid's geometry, which applies in the absence of gravity, and (b) Schwarzschild's geometry, which applies near a massive spherical object, whose boundary is shown by the shaded area. In both cases, time intervals are measured by the observers $A$ and $B$, whose world lines are shown by $a$ and $b$, respectively. The dashed lines are tracks of light signals emitted by $B$ at regular intervals.

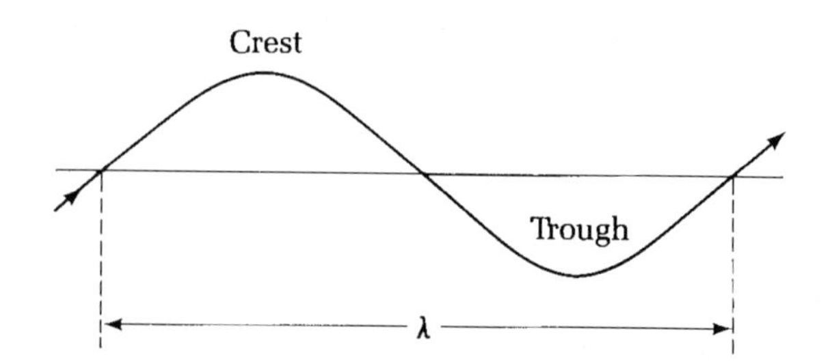


Figure 7-5. The profile of a typical wave. From the level position, the wave rises to the maximum height, the crest, then passes through the level position down to the minimum height, the trough, and then rises back to the level position. The wavelength $\lambda$ measures the distance between the initial and the final level positions. It also measures the distance between successive crests or troughs of the wave. If in one second $v$ wave crests pass a given point, the wave will have advanced through a total distance $v \times \lambda$. This is the speed of the wave.

*A*. These same wave crests would be received by *A* in the interval $(1 + z)$ seconds. In other words, the frequency of waves received by *A* is reduced to $v/(1 + z)$.

There is a simple relation (illustrated in Figure 7-5) between the frequency $v$ and wavelength $\lambda$ of a light wave:

$$\begin{array}{ccccc} \text{frequency} & \times & \text{wavelength} & = & \text{speed of light}, \\ v & \times & \lambda & = & c. \end{array}$$

In Figure 7-4b, the frequency of the light wave is *reduced* by the factor $(1 + z)$ in going from *B* to *A*. The wavelength is therefore *increased* by the factor $(1 + z)$. That is, if $\lambda_B$ and $\lambda_A$ are the wavelengths of the light wave as measured by *B* and *A*, respectively, we have

$$\lambda_A = (1 + z) \times \lambda_B.$$

If *B* were emitting a whole spectrum of visible light (as a star does), *A* would receive the spectrum with all its wavelengths systematically increased by this factor $(1 + z)$. Of all the colours that make up the visible spectrum, red light has the largest wavelength. Therefore, relative to *B*'s spectrum, *A*'s spectrum will have shifted toward the red end. Hence, $z$ is named the *redshift*. Since this effect has been caused by the gravity of the massive object, it is known as the *gravitational redshift*.

This effect is very small for the Sun. In general, for weak gravity, we have a simple formula for $z$:

$$z = \frac{GM}{c^2 R}.$$

This is the redshift of light leaving the surface of an object of mass $M$ and radius $R$. For the Sun, $z$ is as small as two parts in a million.

For some other stars, this effect is somewhat larger. The best case, from a practical point of view, is that of *a white dwarf* star. Here $z$ lies in the range from $10^{-5}$ to $10^{-3}$. For example, for the white dwarf Sirius B, the gravitational redshift is as high as 300 parts in a million. Small though it is compared to 1, the gravitational redshift is an indication of the effect of non-Euclidean geometry on time measurements.

## THE EVENT HORIZON

We now come to the more dramatic aspects of this redshift, when the gravitational effects are strong. Let us do a thought experiment on a star undergoing gravitational collapse. We station an observer $B$ on the star's surface with instructions to send out signals every second. We station another observer $A$ far away from the star and along the radially outward direction from $B$. The situation, illustrated in Figure 7-6, is similar to that of the gravitational redshift from the surface of a white dwarf star. The only difference between the white dwarf and the present collapsing star is that the surface on which $B$ is situated is collapsing. As the star collapses, $B$ successively encounters an increasing force of gravity.

From our formula for the gravitational redshift, we find that, as the star contracts, its radius $R$ *decreases* and the value of the redshift $z$ *increases*. There is, however, one important difference from our previous example of the white dwarf. This formula is valid in situations of *weak* gravity. Our collapsing star with its decreasing radius will soon encounter *strong* effects of gravity. In such cases, this formula is changed to a rather complicated form. The curve in Figure 7-7 illustrates how the redshift changes with the radius $R$ according to this formula. The redshift actually observed by $A$ will not only be of gravitational origin, but will also have an additional

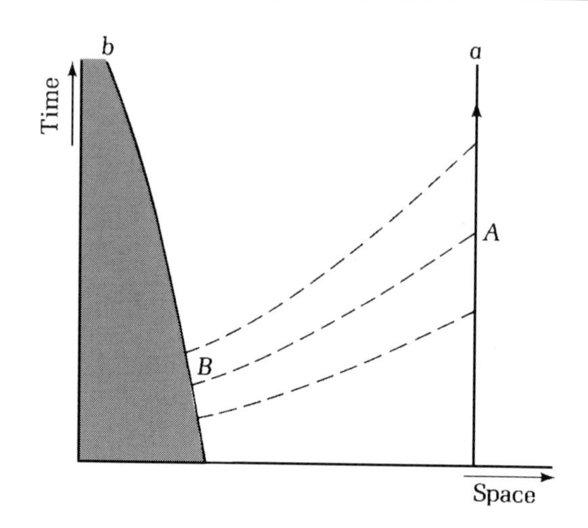


Figure 7-6. Signal propagation between $A$ and $B$ in the case where $B$ is located on the surface of a collapsing object (*shaded area*). Although $B$ sends signals at regular intervals, they reach $A$ at progressively longer intervals as measured by $A$'s clock.

contribution because $B$ is not stationary with respect to $A$ *but is moving away from A*. This latter contribution is commonly known as the *Doppler effect*.

The Doppler effect tells us that whenever a source of light moves away from an observer, the light from the source as seen by the observer is redshifted. The effect is observed for sources of sound also. In Figure 7-8, we see an observer standing at a railway crossing while a train passes by. The whistle of the engine appears to this observer to be *shrill* while the train approaches and *flat* as the train recedes. The shrillness and flatness in this example indicate increase and decrease in the frequency of sound waves. The corresponding effects for light waves are known as the *blueshift* and *redshift*. In our example of the collapsing star, $B$'s signals are redshifted because of both the Doppler effect and gravity.

The curve of Figure 7-7 shoots upward as $R$ approaches the quantity

$$R_S = \frac{2GM}{c^2},$$

indicating that the redshift becomes infinitely large at $R = R_S$. This

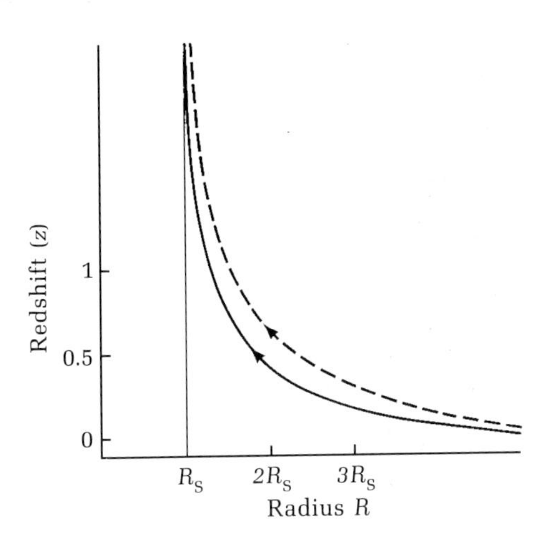


Figure 7-7. A schematic representation of how the redshift from the surface of a collapsing star increases as the star contracts. The curve describes the redshift increase as the star contracts *slowly*. $R_S$ is the Schwarzschild radius of the star. The redshift becomes infinite as $R$, the radius of the star, approaches $R_S$.

significance of $R_S$ is related to the work of Karl Schwarzschild, and for this reason $R_S$ is often called the *Schwarzschild radius*. The subscript S denotes this. Schwarzschild had solved Einstein's equations of relativity back in 1916. The significance of this solution, however, became clear much later, and we will discuss it now in the context of the shrinking star.

As $R$ approaches $R_S$, $z$ becomes infinite. To understand the meaning of this statement, recall that the factor $(1 + z)$ denotes the ratio of the rates at which the clocks of $A$ and $B$ run. The time interval between successive signals appears as one second to $B$ and as $(1 + z)$ seconds to $A$. As $(1 + z)$ increases, $A$ has to wait longer to get the next signal from $B$. When $(1 + z)$ becomes infinite, $A$ will have to wait *forever* for the next signal from $B$! In other words, as $B$ crosses the sphere of Schwarzschild radius $R = R_S$, *no* signals from $B$ will ever reach $A$. This is illustrated in Figure 7-9a.

Interpreted this way, the surface of radius $R_S$ constitutes an *event horizon*. No events taking place within the event horizon will ever be observed by an outside observer like $A$. The analogy here is with the horizon on the Earth. As shown in Figure 7-9b, the curvature

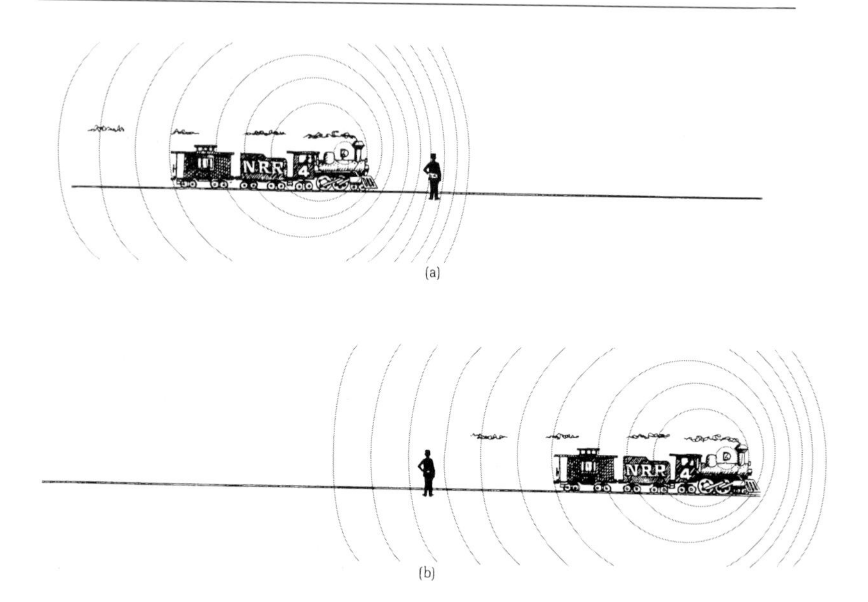


Figure 7-8. The Doppler effect. When the train is approaching the observer, the sound waves emitted by the whistle of the engine are closer together. This is shown in the figure by the narrow spacing of successive wavefronts as the waves move from the engine to the observer. The opposite effect takes place as the engine recedes from the observer.

of the surface of the Earth limits the range of its visibility from any given height. A ship observed on the high seas from a lighthouse disappears from view when its distance exceeds this range. Just as the ship is not visible beyond the Earth's horizon, so is $B$ not visible once it crosses the event horizon.

The situation we have just discussed has an amusing analogy in everyday life. Imagine $A$ to be an applicant for some financial grant from a highly bureaucratic government agency. $B$ is the bureaucrat handling $A$'s correspondence. The time scales of $A$ and $B$ are, however, different. Whereas $A$ expects a reply to his query within a day, $B$ may take several days to oblige with an answer. To $B$, whose normal existence is tuned to the red tape in his office, this delay would appear quite normal. We may call this phenomenon *bureaucratic redshift*. In a survey conducted by the administration of a state in India, it was found that a typical letter waits in the files of

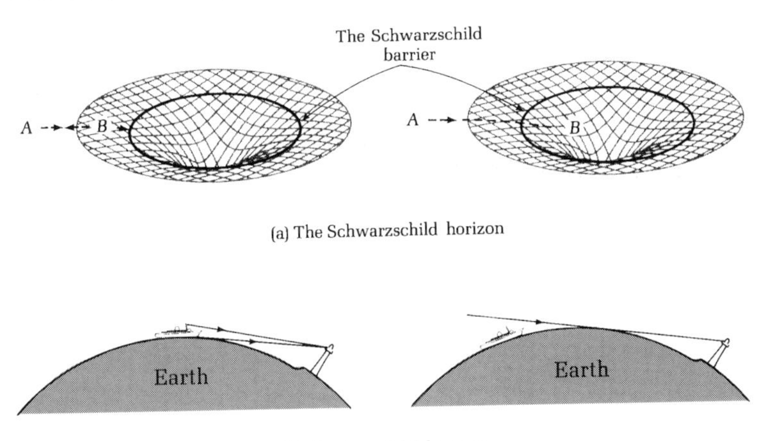


Figure 7-9. Two horizons. (a) In the figure at left, $A$ and $B$ are able to communicate with each other along the two-way track shown with arrows. $B$, however, is moving away from $A$ into a region of increasingly curved spacetime. In the figure at right, $B$ has crossed the Schwarzschild barrier (shown by a thick closed curve). Once this happens, no signal from $B$ reaches $A$. This barrier is the Schwarzschild horizon. The word 'horizon' is used in analogy with the terrestrial horizon shown in (b), where the range of visibility at the lighthouse is limited by the curvature of the Earth's surface.

the bureaucrats for an average period of 288 days before receiving a reply! If we set our time scale ratio of $(1 + z)$ equal to 288, the bureaucratic redshift in this case equals 287.

But to return to our collapsing star: its gravitational redshift exceeds even this high value as it collapses further. When the surface of the object crosses the event horizon (that is, when its redshift becomes infinite), it becomes a black hole.

## A BLACK HOLE HAS NO HAIR!

Can the outside observer $A$ say when $B$ has disappeared from view? Strictly speaking, $A$ may wait *forever* but still not actually be able to say, 'Now I know that the star has become a black hole.' The signal emitted by $B$ just at the moment of crossing the event horizon is destined never to reach $A$. Information about the history of $B$ *prior to* this crucial instant is, however, accessible to $A$ provided $A$ waits long enough.

Astronomers and astrophysicists often say that there is a black hole in such and such an astronomical system. How can these statements be justified in view of what we have just said? Surely, any person on the Earth viewing an astronomical event is like observer $A$. Never during his or her lifetime (however long that may be) can he or she claim that a black hole has been formed. In this exact sense, the claims made in the popular or technical literature about the existence of black holes are false.

In an approximate sense, however, such claims may be justified. Even before the star has actually crossed the event horizon, it may for all practical purposes become invisible to the external observer $A$ because the gravitational redshift also acts in a way that drastically lowers the luminosity of the star.

We have seen how the frequency of light emitted from the surface of the star is reduced by the factor $(1 + z)$ by the time it reaches $A$. The energy contained in a beam of light of a given frequency can be measured by first counting the number of *photons* in the beam.

What is a photon? At the beginning of this century, thanks to the ideas of Max Planck and Albert Einstein, the dual nature of light became apparent (see Figure 7-10). Although we have so far been thinking of light as a wave, its behaviour in some microscopic situations is best explained by imagining that it is made of tiny *packets of energy*, called *photons*. A photon is thus an elementary particle that carries light in the form of an energy packet. Each packet has a certain energy, calculated by multiplying its frequency by a universal constant known as Planck's constant (usually denoted by $h$). So, as the beam of light travels from $B$ to $A$, *its energy is reduced by the same factor by which its frequency is reduced.* This reduction factor $(1 + z)$ becomes very large as the star's surface approaches the event horizon.

Because of this drastic reduction in its energy output, the star appears *almost black* even before it has reached the event horizon. Since all astronomical detecting instruments have thresholds of energy flux below which they cannot detect the sources of radiation, as soon as the collapsing star's energy flux drops below those thresholds, it will cease to be seen, even though it has not yet crossed the

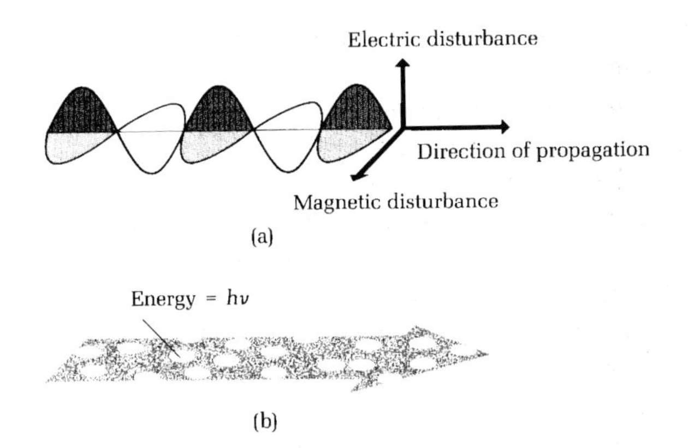


Figure 7-10. Light has a dual nature. In (a), light is shown as a wave with electric and magnetic disturbances propagating in wave patterns in mutually perpendicular directions. The positive disturbances are shaded; their maximum values correspond to wave crests. The number of crests per second equals the frequency ($\nu$) of light. In (b), light is shown as being made of packets of energy called photons. A photon of light of frequency $\nu$ has energy $h \times \nu$, where $h$ is known as Planck's constant.

event horizon. In this approximate sense, the star may be said to have become a black hole. It is in this sense that we will henceforth use the statement that the star has become a black hole.

Even if a star has become a black hole in this way – that is, even if its radiation is too faint to be seen – its existence can be inferred by other indirect means. In particular, its gravitational effects will continue to be present. If the star had planets orbiting around it, these planets would continue to go around in their old orbits. In Chapter 3, we saw how we can estimate the mass of the Sun from information about the size and the period of the Earth's orbit. By similar techniques, we can in principle estimate the mass of a star even if it has become a black hole.

What other information can we get about a black hole? Precious little, if some of the exploratory work in this field turns out to be correct! In our example of the gravitational collapse of the star, we assumed the star to be a perfect sphere. The collapse of a spherical object can be easily described by Einstein's equations, which have explicit solutions. However, if the collapsing object has a highly irregular shape, if it is rotating as well as collapsing, or if it has electric charges and currents with their associated magnetic fields, then the problem of solving the equations of collapse becomes impossibly difficult. That is, given the initial stages of such a collapse, we cannot find the detailed sequence of stages through which the object in question becomes a black hole.

The exploratory work referred to above describes a somewhat limited range of initial stages. It is concerned with the situation where the departure from the spherical state is rather small. The conclusion based on such work, due largely to Richard Price, may be described as follows.

In Figure 7-11, we see the initial stages of collapse of an irregularly shaped object. All the different types of irregularity that we referred to earlier are present. Yet, in the final stage, when the object has become a black hole, what information is left for the outside observer *A*? To *A*, the black hole will appear to have a mass, some angular momentum, and some electric charge. By designing suitable experiments, *A* can measure these three quantities but no more! Information about all other aspects that were present in the initial stages will have disappeared. John Wheeler has described this situation by his oft-quoted remark, 'The black hole has no hair!'

Because so little information about the black hole remains accessible to the outside observer, the detection of a black hole has to depend largely on indirect evidence. This has not deterred theoreticians, however, from putting forward ingenious scenarios in which black holes have important roles to play. We will concern ourselves with a few examples in the following chapter.

## SPACETIME SINGULARITY

We end this chapter with a brief discussion of what happens to *B* after this observer has crossed the event horizon. We must remind ourselves that, although *A* has to wait *forever* to witness *B*'s crossing of the horizon, *B* of course does not notice this slow passage of time. By *B*'s clock, the crossing of the event horizon takes very little time. In fact, by *B*'s clock (if *B* were on the surface of the Sun!), the time taken for the Sun to undergo a gravitational collapse in the unlikely event of a sudden withdrawal of its internal pressures is only about 29 minutes, right from its present size to the final pointlike state.

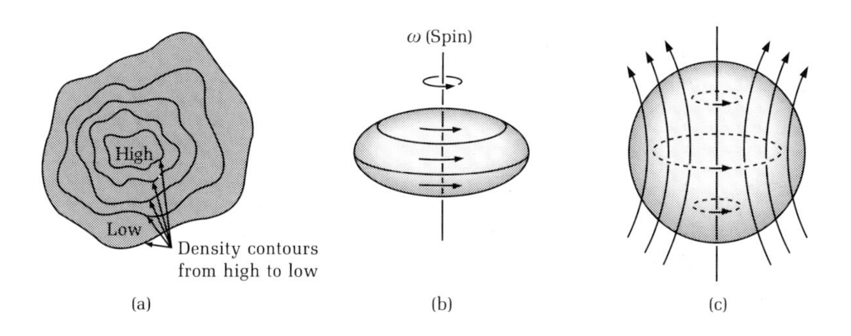


Figure 7-11. This figure illustrates what details of a collapsing body can be observed by an outside observer as the body approaches the event horizon. (a) Even if the collapsing body is irregular in shape, the surviving information tells the outside observer *nothing* about these irregularities. He can only measure its mass. (b) A spinning body has angular momentum. This can be determined by the outside observer even when the body has become a black hole. (c) Even though the body may have electric currents (*dashed lines*) and magnetic fields (*solid lines*), these details are lost. The outside observer only measures the net electric charge in the body. (From *The Physics–Astronomy Frontier* by F. Hoyle and J. V. Narlikar. Copyright © 1980. W. H. Freeman and Company.)

There are, however, other effects that *B* will begin to notice with the collapse of the star – effects that will prove to be highly uncomfortable. In Chapter 6, we described tidal forces. These forces (see Figure 7-12) will tend to stretch *B* in the radial direction. How big are these effects? For a star 3 times as massive as the Sun, the tidal tension becomes 100 times the atmospheric pressure on the Earth's surface by the time the star's radius is 40 times the Schwarzschild radius. No human being can withstand this disruptive tension. And by the time the Schwarzschild radius is reached, this tension will have increased 64 000 times!

Even if *B* somehow summons incredible resources to survive tidal disruption, there is worse to follow. As the star contracts inside the event horizon, the geometry of spacetime around *B* will become increasingly peculiar. (The growing tidal force is just one manifestation of this peculiarity.) The climax is reached when the star shrinks to a point. This is 'The End' for the star as well as for *B*.

For not only does the star have infinite density, but the spacetime around it becomes infinitely curved. Mathematicians describe this

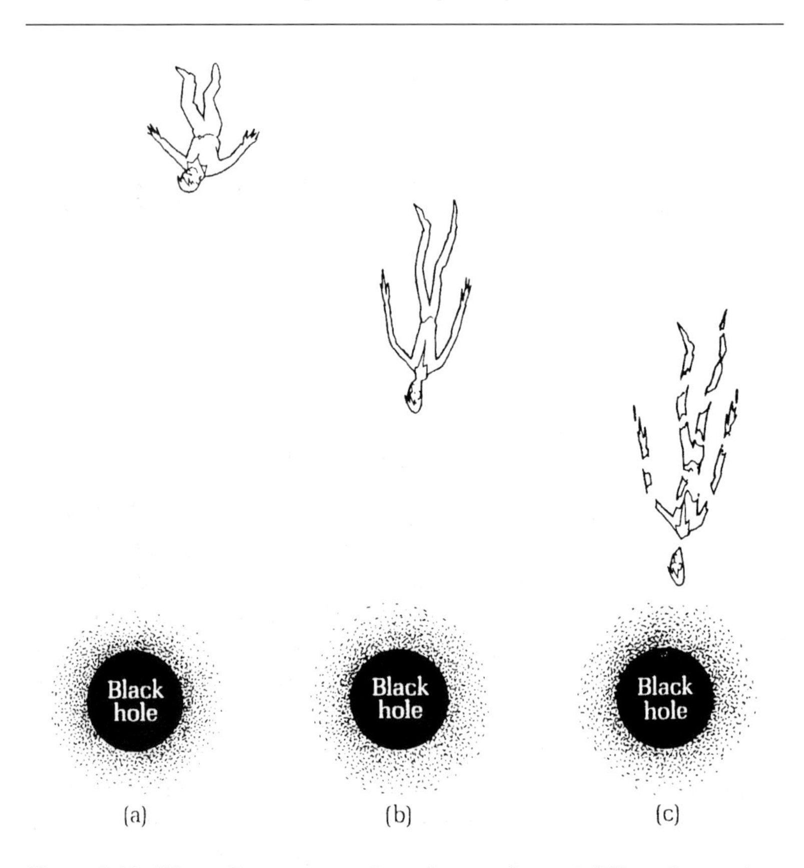


Figure 7-12. The unfortunate experience in store for an infalling observer is to be torn apart by the huge tidal forces of the collapsing object as shown here for a man falling freely toward a black hole. In (a) the man is far away from the black hole and hardly notices the black hole's tidal force. In (b) he is near enough to be stretched by the tidal force, while in (c) as he approaches still closer to the black hole he is torn apart by the tidal force.

as a state of *singularity*. All mathematical equations break down at this state. The laws of physics cease to operate at the instant of singularity. Beyond this instant, the future is unpredictable. In fact our well known concepts of space and time cease to have any meaning!

# 8

# *Cosmic energy machines*

## THE COSMIC ENERGY PROBLEM

We here on Earth are constantly reminded by experts that with advancing technology our energy needs are growing, and that we need to worry about stocks of oil, coal, nuclear fuel, etc. that are needed to generate energy to meet these demands all over the world. How long will the supplies last? Can we extend that period by conserving energy? If so, how? These questions are being debated by experts and lay people alike.

Astronomers face the 'energy problem' in their investigations of cosmic sources of radiation. The age-old problem, where the Sun gets its energy to shine so brightly and steadily, has been solved. In Chapter 4 we saw that the key to solar energy lies in the nuclear fusion going on in the central core of the Sun.

But in the 1950s new problems with far greater magnitude began to confront the astronomers. The radio astronomers began to find sources of radio emission whose total energy reservoir exceeded that of the Sun by several billion. Where did the source of this energy lie? The problem was exacerbated in the early 1960s with the discovery of quasi-stellar sources, commonly called quasars. Initially mistaken for stars, quasars turned out to be far more energetic, and far more dramatic in spending their energy.

A typical quasar radiates in visible light as much as a galaxy of hundred billion stars. It also radiates in X-rays and possibly other wavebands. What is more, the main core of a quasar where the energy reservoir is concentrated may be no greater than the size of the solar system! The idea of nuclear energy generation based on star models does not work under these conditions. Something more dramatic is needed.

The key to this problem was provided in 1963 by Fred Hoyle and William Fowler, at the time the first two quasars with catalogue names 3C 273 and 3C 48 were discovered. Hoyle and Fowler suggested that the compact nature of these objects implied the dominance of the force of gravity in the scenario governing their overall equilibrium. As we saw in the last chapter, gravity begins to dominate as the object contracts. The phenomenon of gravitational collapse is the outcome of complete dominance by gravity over other forces. So Hoyle and Fowler's idea was to employ the strong gravity of a compact and very massive object to generate energy at the expense of its huge gravitational energy reservoir.

In 1963 the black hole had not entered the theoretical scene, so the Hoyle–Fowler idea may be seen as a precursor to it. The black-hole concept began to consolidate in the 1970s and people began to appreciate the role a massive black hole can play as an energy machine. We will first concentrate our attention on black holes, drawing on their strange properties, which we discussed in the previous chapter. Later we will bring in the concept of another object, the *white hole*, that is the exact opposite of a black hole.

## SPINNING BLACK HOLES

We begin with a discussion of the physical properties of spinning black holes. In Chapter 7 we explained how black holes could form as a result of the gravitational collapse of a massive star. There we assumed that the star was always spherical during its contraction. The black hole that forms out of such a contraction is also spherical. The geometry outside the spherical black hole is that of Schwarzschild. For this reason, it is called the *Schwarzschild black hole*. It is entirely characterized by its mass $M$.

It may happen, however, that the collapsing star is spinning about an axis. The laws of dynamics then tell us that the angular momentum of the star must remain constant as it collapses. As a result of this rule, the star will spin faster and faster as it contracts. It may also happen that this process could lead to a disruption of the star and altogether prevent its reaching the black-hole stage. However, if this disruption does not take place, then the resulting black hole will also possess angular momentum. (See Chapter 1 for a definition of angular momentum.)

The detailed process of how – if at all – a spinning object attains black-hole status is not well understood. Therefore, we cannot say definitely under what conditions a spinning black hole is formed. Nevertheless, we can say what a spinning black hole is like once it has formed. In 1963, Roy Kerr gave a description of the geometry of spacetime in the empty region outside a spinning object symmetrically shaped about its axis of spin. Kerr's solution of the problem has greatly added to our understanding of the spinning black hole, which is sometimes referred to as the *Kerr black hole*.

Figure 8-1 shows two sections of a Kerr black hole. In 8-1a we have a meridian section, that is, a section through the axis of rotation. The inner circle (solid curve) shows a section of the spherical *event horizon*. The outer broken curve denotes the section of the boundary of what is known as the *ergosphere*. We will presently see the reason for this name. In 8-1b we see a section of the black hole at a given latitude. The inner and outer circles, whose common centre lies on the axis of rotation, denote respectively the sections of the event horizon and the ergosphere.

Imagine two observers $B_1$ and $B_2$. As shown in Figure 8-1b, $B_2$ lies outside the ergosphere while $B_1$ lies inside it but outside the event horizon. Now, our statement that the black hole spins about an axis has meaning only in relation to some background – say, the background provided by the distant stars. Suppose $B_1$ and $B_2$, who can see this background, want to maintain themselves at rest relative to this background. Can they achieve this?

A rough analogy of the situation is seen in the example of an aircraft flying above the rotating Earth. An aircraft that flies above the surface of the Earth is carried along the direction in which the Earth rotates, that is, from west to east. If this were not so, it would be a simple matter to travel west. The aircraft would go up, stay in one place, and come down when the destination appeared below.

In the case of the Kerr black hole, both $B_1$ and $B_2$ are carried along the direction in which the black hole rotates. To stay in one place, both $B_1$ and $B_2$ would have to apply extra force – say, by firing rockets – to counter this tendency of the black hole to carry them

and lower angular momentum than before. Thus, it is possible in principle for an advanced civilization to operate the spinning black hole as a powerhouse by simply firing chunks of matter at the black hole and receiving back pieces with greater energy than the energy spent on the ingoing chunks.

Ingenious though the Penrose process is, it is highly esoteric. So far, no astronomers have been able to 'cook up' a scenario based on the Penrose 'recipe'. There are, however, plenty of examples in high-energy astrophysics where rotating black holes have been suggested as sources of energy. In the examples discussed below, it will become apparent that the main properties of black holes utilized in the production of energy are their concentration of a lot of matter in a small region of space and their angular momentum.

## CYGNUS X-1

In Chapter 6, we described the tidal effects that arise in a binary star system. Let us follow that scenario further.

Recall that, in the binary system shown in Figure 6-8, two stars $A$ and $B$ go around each other in elliptical orbits. Star $B$ is compact, while star $A$ is extended. Tidal effects become important when $A$ extends beyond its *Roche lobe*. The surface matter from $A$ is then pulled by its companion $B$. What happens to this pulled-out material?

Figure 8-3 illustrates what happens in a schematic way. If the double star system were not rotating, the material pulled out from $A$ would have gone straight toward $B$. Because of rotation, however, this material does not go straight away to $B$ but is made to rotate around $B$. As shown in Figure 8-3, the matter from $A$ goes around and around $B$ mostly in their common equatorial plane, where the tidal effect is largest. It finally falls onto $B$. This process results in the formation of a disk of infalling matter around $B$.

In the 1940s, Hermann Bondi, Raymond Lyttleton, and Fred Hoyle considered various situations where a star can *accrete* matter from interstellar space as it moves through it. The strong gravitational pull that the star exerts on the surrounding matter explains the accretion. The problems considered by these astrophysicists were largely of spherical accretion – that is, accretion from all di-

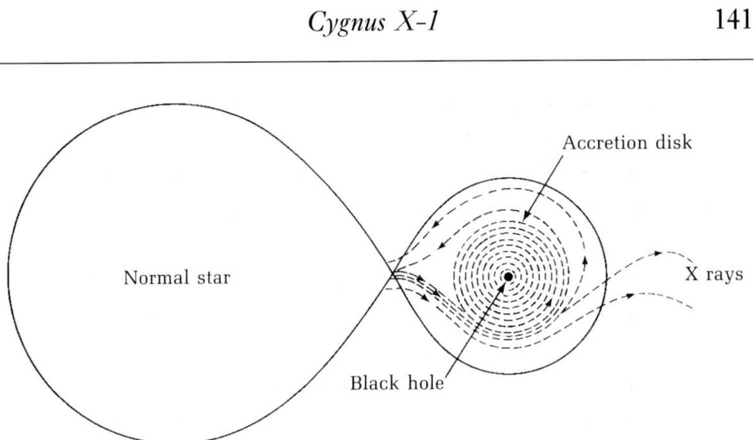


Figure 8-3. The formation of an accretion disk around a black hole in a binary star system. The normal star is $A$ and its black hole companion is $B$ (see text). (From *The Physics–Astronomy Frontier* by F. Hoyle and J.V. Narlikar. Copyright © 1980. W.H. Freeman and Company.)

rections onto a spherical star. Bondi's classic paper of 1951 formed the starting point of modern accretion theories.

We now consider the same concept of accretion in the context of a binary star system. The difference here is that the accretion is in a disk rather than from all directions. The size of the disk is determined by various parameters, such as the rate of accretion, the rate of infall onto star $B$, the mass of star $B$, and so forth. The size of the disk can fluctuate if the accretion rate varies. The interesting effect of such a disk is, however, that the matter in it becomes *hot* as a result of friction, and this hot matter radiates. The radiation frequency depends, among other things, on the temperature of the disk. Calculations show that this radiation should be largely in the form of X-rays.

The main role in this picture is, of course, played by the star $B$. For a radiating accretion disk to form around it, the star $B$ must be a very compact star. In the early 1960s, neutron stars were discussed anew by astrophysicists; in the late 1960s, black holes began to gain prominence. Thus, both neutron stars and black holes were recognized as good candidates for star $B$ in this accretion-disk mechanism.

Along with this theoretical development, the progress of space technology made it possible for astronomers to observe cosmic X-ray sources. Since X-rays from outer space are absorbed by the Earth's atmosphere, their observation becomes feasible only if we can set up detector instruments above the atmosphere. Such attempts were made in the 1960s. A major advance in X-ray astronomy was the survey conducted by the Uhuru satellite* launched on 12 December 1970. This survey revealed the existence of many X-ray sources, among which was one in the constellation of Cygnus. This source, now known as Cygnus X-1, is the best example of the binary star X-ray emission described above.

The binary star system at the location of Cygnus X-1 consists of a supergiant star $A$ and a compact companion $B$. The star $A$, known as HDE 226868 in the *Henry Draper Catalogue*, is the only visible star for this binary system. Its mass is at least $20M_{\odot}$, and it appears to be orbiting about its companion, taking about $5\frac{1}{2}$ days to complete one orbit. The companion $B$ cannot be seen, but its existence can be inferred. It is this object that has caused so much excitement among black-hole astrophysicists.

Detailed observations of the orbital parameters of this double-star system suggest that the mass of $B$ must be at least $8M_{\odot}$. This high value far exceeds the theoretical upper limit on the mass of a neutron star, a limit that is known not to exceed $3M_{\odot}$ and is probably closer to $2M_{\odot}$. This result, therefore, eliminates $B$ as a standard neutron star. What else can it be? The alternative is a black hole. For this reason, Cygnus X-1 is often cited as the best-known observational evidence for the existence of a black hole.

This evidence for a black hole is highly suggestive but still circumstantial. Thus, a sceptic may still feel that an alternative scenario not involving such an esoteric object as a black hole might eventually be found.

However, if the black-hole interpretation of Cygnus X-1 is speculative, the other current alternative interpretations suggested from time to time are even more so. In any case, the example of Cygnus X-1 has prompted many black-hole astrophysicists to be more daring and to consider the applications of the accretion-disk scenarios on even larger scales. We consider next a few examples of such applications.

* *Uhuru* means 'freedom' in Swahili, the language of Kenya, where the satellite was launched on the anniversary of Kenya's Independence Day.

## SUPERMASSIVE BLACK HOLES

Considerations of stellar evolution led us to the concept of black holes with masses that are a few times the mass of the Sun. The black hole believed to exist in the double-star system giving rise to Cygnus X-1 has a mass of about $8M_{\odot}$. We now consider black holes far more massive than those arising as the end states of stars. These black holes, known commonly as *supermassive black holes*, are thousands to billions of times more massive than the Sun, and they are conjectured to exist in galactic nuclei and quasars. Their role in these astronomical systems is to generate energy.

In 1966 Fred Hoyle and I had suggested that highly collapsed massive objects as much as a billion times as massive as the Sun might form the nuclei of elliptical galaxies. The name 'black hole' was not in vogue then but the idea that a dense mass at the centre might gravitationally control the size and shape of an entire galaxy was considered rather unusual in those days. Within two decades, however, observations began to suggest the existence of compact massive galactic nuclei. The first such spectacular case is of the galaxy M87 (87th in the *Messier Catalogue*), shown in Figure 8-4.

From early observations in 1978 the mass of the black hole inside M87 came out to be as high as $5 \times 10^{9} M_{\odot}$! Speculations apart, what evidence is there in support of such claims?

First, let us try to understand what these supermassive black holes are expected to do. In general, if such black holes do form, they must arise from the gravitational collapse of a collection of stars. Such a collection of stars has an overall angular momentum that survives even when the black hole is formed. The black hole, therefore, is of the spinning (Kerr) type. And the spinning black hole tends not only to attract the surrounding matter but also to move this matter around the axis of rotation. The effect is largest in the equatorial plane of the black hole, where this surrounding matter forms an accretion disk. This accretion disk is qualitatively similar to that which forms around a black hole in a binary system, although in actual size and mass it is enormously bigger. The black hole may generate energy through the heating of this disk as in the binary system.

With regard to M87, two teams of astronomers using different observational techniques came to the conclusion in 1978 that a black

Figure 8-4. A photograph of the galaxy M87. Notice the jet coming out from the centre. Careful studies of the central region of this galaxy suggest that there may be a black hole 5 billion times as massive as the Sun in the nucleus of the galaxy. (Courtesy of Palomar/California Institute of Technology.)

hole of about $5 \times 10^9 M_\odot$ in the nucleus of the galaxy might best account for their observations. One team of astronomers from the Hale Observatory and the Jet Propulsion Laboratory measured the brightness of visual light from all over the galaxy. Although such brightness measurements had been made previously for many galaxies, the sensitivity achieved in the 1978 results with the new invention of the charge coupled device (CCD) was far higher than before, which enabled the astronomers to look at the nuclear region of M87 more carefully. They found that the brightness profile of the galaxy rose sharply toward the centre instead of flattening out, as shown in Figure 8-5. Such a rise in luminosity indicates a dense conglomeration of stars in the nuclear region, which in turn points to a strong gravitating object that keeps the stars huddled together near the centre. What could this object be? After examining several alternatives, the observers found the black-hole interpretation to be the most suitable one.

The other team of astronomers (from the Hale Observatory, the Kitt Peak National Observatory, University College, London, and

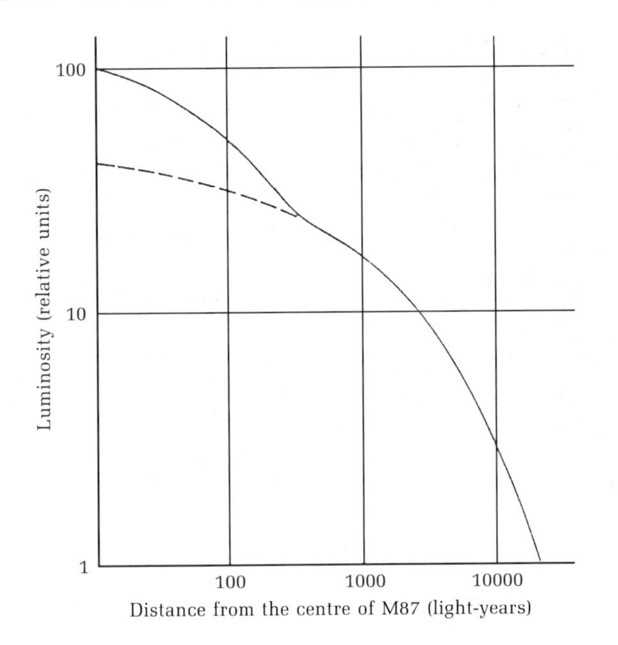


Figure 8-5. The luminosity profile of M87. Notice that, instead of reaching a flat plateau toward the centre (as shown by the dashed curve), the luminosity shows a sharp rise, which points to some compact energy source in the centre of M87. A favourite explanation for this energy source is that it is a black hole. The graph uses a logarithmic scale both for luminosity and distance. The rise in luminosity from the outer end of the profile toward the peak at the centre is by a factor of 100.

the University of Victoria) examined the nuclear region spectroscopically. Their measurements give indications of the extent of spread in the speeds of stars in the nuclear region. The larger-than-expected spread in speeds implied enhanced dynamical activity that might have been caused by a supermassive black hole in the nucleus.

However, there are now some doubts as to whether the increased dynamical activity can be definitely ascribed to a black-hole, as the rise is not as sharp as expected.

The rapid advance in observing techniques using the CCD as well as corresponding improvements in other areas of astronomy have led to several cases of direct or indirect evidence for a high concentration of matter in the nuclei of very active galaxies (that is, galaxies showing evidence of explosive activity), quasars, and even relatively normal and supposedly quiet galaxies like our Milky Way. The indirect evidence is indicative of high and rising dynamical activity toward the centre, while the more direct evidence is usually of a growing intensity of radiation toward the centre. We saw instances of both kinds in the early example of M87.

Look at Figure 8-6, for example, which is a radio picture of a couple of jets coming out in opposite directions. There is a central galaxy but the jets themselves are believed to collide against an intergalactic medium and give rise to lobes of radio emission. The energy involved in the process may be as large as $10^{53}$ joules, enough to keep the Sun shining at the present rate for a *billion billion years*! Yet the source of the jets must lie in the tiny central region, the compact nucleus of the galaxy. The belief is that a supermassive black hole at the centre pulls in matter from the surroundings into a disk and somehow diverts its infall energy along the two outward jets. The gravitational energy source of a billion-solar-mass black hole may be just about sufficient to account for the phenomenon, although a detailed dynamical model of the scenario is still missing.

What evidence do we have for an accretion disk? Just as the existence of a black hole is to be inferred, so is the existence of its surrounding disk, which is barely ten to a hundred times larger. For example, a black hole of a billion solar masses will have an accretion disk no larger than, say, three hundred billion kilometres in diameter. For us terrestrials this distance looks huge; but measured in light-years it is just about 3% of a light-year. Sophisticated though today's observing techniques are compared to two decades ago, they are still inadequate to delineate an object of this size at a typical distance of, say, ten million light years. This is like looking for a needle at a distance of a thousand kilometres. Even though the needle shines (as the accretion disk would), you will agree that this is going to be a very difficult task! For this reason most evidence is of an indirect kind: all it tells us is that the large-scale features of the source are consistent with there being a black hole with an accretion disk.

In May 1994 the Hubble Space Telescope demonstrated how much more clearly it can image distant astronomical objects; when it showed a clear image of a disklike structure around the centre of M87. Velocity measurements of the gaseous disk showed that it is rotating. The mass of the black hole is estimated as being

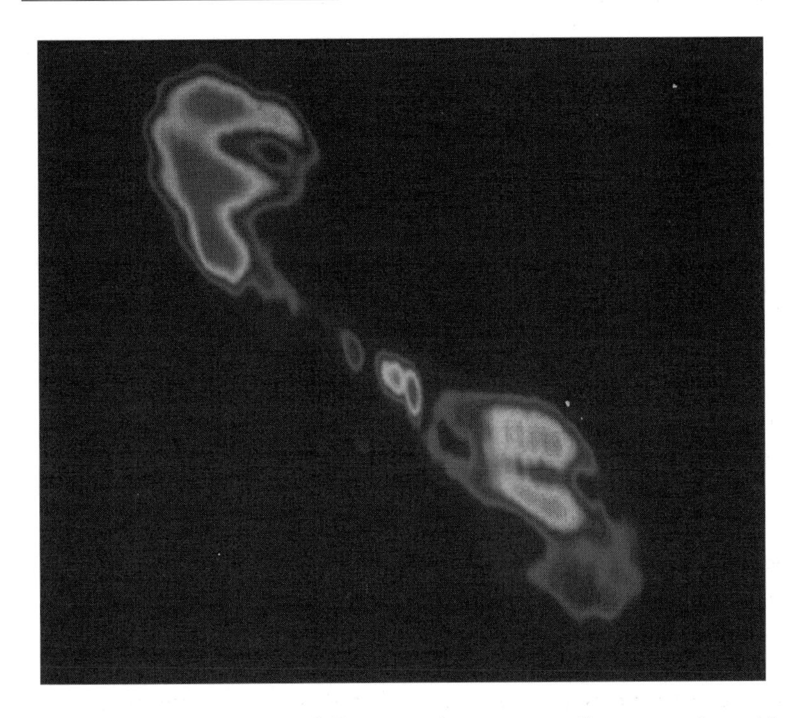

Figure 8-6. The radio map of Centaurus A, a strong radio source taken with the Very Large Array in New Mexico, USA. (Courtesy of the National Radio Astronomy Observatory, USA.)

three billion solar masses, while the radius of the disk is at least 60 light-years. Also, the disk has a temperature of about ten thousand degrees Celsius.

This certainly is the best evidence to date for the black-hole hypothesis. However, a sceptic may still argue that at 60 light-years the size of the gaseous disk is far larger than the theoretical value of a few percent of a light-year. Thus, what is being seen is not the accretion disk.

There are, of course, a few leading astronomers and physicists who feel that the credence given to black holes is somewhat exaggerated. Since black holes cannot be directly observed (by definition!), their existence must be *inferred* from indirect measurements. These examples illustrate the maxim of Sherlock Holmes, the great fictional detective of Sir Arthur Conan Doyle, which may be paraphrased as follows: 'When all other plausible astrophysical explanations fail, the black-hole interpretation, however bizarre it may seem, must represent the truth.'

Some astronomers feel that all other plausible astrophysical explanations have *not* been fully explored, and therefore the inference of the existence of black holes rests on insufficient evidence. This scepticism about astronomical black holes, which arises in part from their bizarre nature, again reminds us of the Black Hole of Calcutta. The description of the Black Hole of Calcutta is so macabre that some historians doubt whether the event took place at all!

## WHITE HOLES

One reason for scepticism of the black-hole scenario is that all evidence in active galactic nuclei, quasars, etc. shows outward, rather than inward, motion of particles and radiation. Thus, the indications, prima facie, are of explosive phenomena rather than implosion. In the black-hole picture it is *presumed* that the inward motion generated by the black hole's attraction is *somehow* converted into outward motion. But the exact process whereby this happens is not specified. Thus, the connection between observations and the black hole is somewhat indirect.

As an alternative to the black-hole scenario, we now describe the so-called *white holes*, which are more directly identified with the explosive events observed. What are white holes?

In our examples of the collapsing star, we had placed one observer, $B$, on the collapsing surface, while the other observer, $A$, was placed far away from the star. The light waves sent from $B$ to $A$ were *redshifted* for two reasons: part of the redshift was of gravitational origin, and part was of Doppler origin.

In a white hole, we have a *time-reversed version* of a collapsing object as seen by $B$. What does time reversal mean? When we observe any phenomenon taking place in nature, we see a sequence of states (of whatever we are observing) ordered according to chronological time. A good example is a movie film, in which we seem to see some event taking place. But the film in fact consists of a succession of still pictures passing rapidly in front of our eyes. If we were to reverse the succession of these still pictures, the film would run backwards. If the sequential order of the original event were reversed, the event seen on such a film would be a new type of event.

Many movie projectors have the capability of running the film backwards, and very often this process leads to hilarious new events being seen on the screen – events in which people walk backwards, food is taken out of the mouth, and the waters of Niagara Falls go up!

The events seen when a movie runs backwards are the *time-reversed* versions of the normal sequence of events. To the physicist, time reversal has a deep significance. The basic laws of physics – whether the laws of gravity, electricity, magnetism, or the interactions in the interior of the atomic nucleus – are all *time-symmetric*. This means that any event taking place according to these laws has a time reversed version that can also take place according to the same laws. The laws *per se* do not make any distinction between the event and its time-reversed version. If these 'strange' events are also possible, why do they not actually take place? It is believed that, in spite of the basic time-symmetry of physical laws, some selection process might be operating in the Universe, a process that allows some events to take place but *not* their time-reversed versions.

In Figure 8-7a, we see some of the different stages during the collapse of a massive spherical star. These correspond to the still frames of a movie film of gravitational collapse. In Figure 8-7b, the order of these stills is reversed, and we see a sequence of states of expansion. The *implosion* into a singularity in Figure 8-7a is now replaced by an *explosion* from a singularity. Notice that the singular state, which was the *final* state of collapse, has now become an *initial* state of explosion.

As Einstein's general theory of relativity is a time-symmetric theory, it permits the explosive situation of Figure 8-7b, just as it permits the implosive situation of Figure 8-7a. This symmetry enables us to talk of white holes as the time-reversed phenomena corresponding to black holes.

Notice, however, that the symmetry of the two situations of Figure 8-7 is manifest in the frame of reference of an observer on the surface – our observer *B* of Chapter 7. To an outside observer like *A*, the black hole and white hole do not appear as time-reversed versions of each other. We will now discover the reason why.

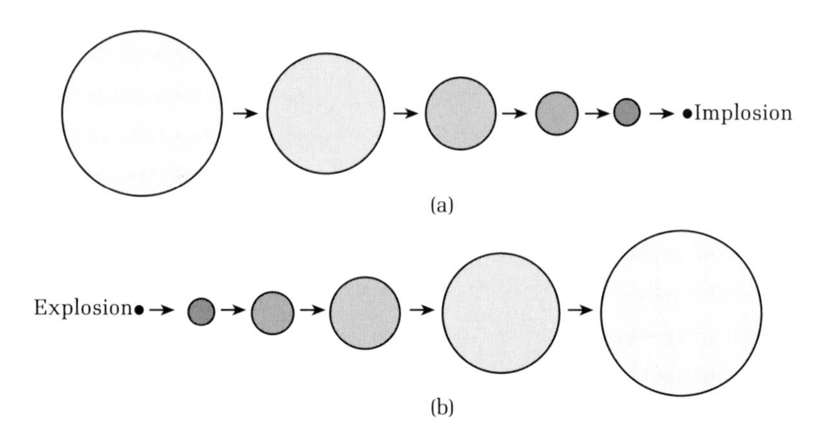


Figure 8-7. In (a) we see in chronological order a few stages in the collapse of a star to a singular state. In (b) the same stages are shown in a time-reversed sequence. The object now appears to explode into existence and expand thereafter. The time coordinate is that of an observer on the surface of the object.

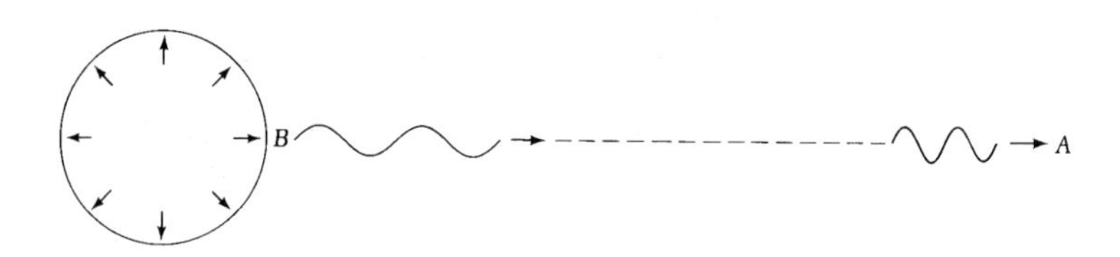


Figure 8-8. Light emitted in the radial direction from $B$ to $A$ can be blueshifted in the early stages of a white hole. The waves emitted by $B$ are shown to be blueshifted by the time they reach $A$; their wavelength has been *reduced* to half the original value.

## BLUESHIFT FROM WHITE HOLES

Let us see what happens to our light-propagation problem in the case of a white hole. In Figure 8-8, we see a radial light wave move from $B$ to $A$. As in the case of collapsing objects, this light travels from a region of strong gravitational effects to a region of weak gravitational effects. Thus, had $B$ been at rest relative to $A$, the light sent out from $B$ to $A$ would have suffered a gravitational *redshift* (as described in Chapter 7).

However, $B$ is not at rest relative to $A$. The surface on which $B$ is located is moving *toward* $A$; hence, the change in the wavelength

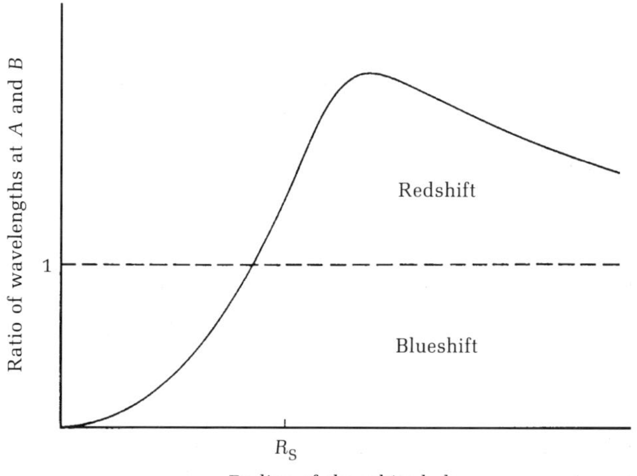


Figure 8-9. The ratio $\lambda_A/\lambda_B$ is very small in the early stages of a white hole, implying the existence of large blueshifts. The ratio increases as the white-hole expansion slows down. It may become 1 either before or after the white hole bursts out of its Schwarzschild radius $R_S$. (In the figure, this happens before the Schwarzschild radius is reached.) Beyond this stage, the blueshift gives way to the redshift of gravity. In late stages of the explosion, the ratio $\lambda_A/\lambda_B$ begins to decline as the gravitational redshift also becomes weaker and weaker.

of light due to the Doppler effect must be considered. The Doppler effect also operated in the collapse problem described in Chapter 7; the difference now is that the Doppler effect produces an *increase* in the frequency and a *decrease* in the wavelength of light as it travels from $B$ to $A$. Applying this result to visible light, we would expect the Doppler effect to shift all wavelengths toward the *blue* end of the spectrum (since blue is the colour near the short-wavelength end).

We therefore have two opposing effects – the gravitational effect tending to increase the wavelength of light from $B$ to $A$ (the redshift), and the Doppler effect doing the reverse (the blueshift). Which of the two effects would win out?

Figure 8-9 illustrates the answer in a qualitative fashion. Here we have plotted the ratio $\lambda_A/\lambda_B$, the ratio of the wavelengths received by $A$ and emitted by $B$, against the size of the white hole at the time of emission. Notice that the dashed straight line corresponds to $\lambda_A = \lambda_B$. Above this line there is redshift, and below it there is blueshift. It is clear that in the early stages of the expansion, when the speed of $B$ toward $A$ is large, the Doppler effect wins out, and the overall effect is of blueshift. In the later stages, the expansion has slowed down and the gravitational redshift is more important than the Doppler blueshift.

Not only does light from the white-hole surface emerge from deep inside the Schwarzschild sphere $R = R_S$, it also emerges with increased energy. Thus, the event horizon that plays the role of a cosmic censor in black-hole physics does not prevent the signals by $B$ on the white-hole surface from leaking outward. To someone only accustomed to black holes, this behaviour of white holes may appear peculiar, but it is nonetheless true.

In 1964, Fred Hoyle, John Faulkner, and I first suggested that this behaviour of white holes makes them ideally suited for generating radiation and particles of high energy. A decade later, Krishna Apparao, Naresh Dadhich, and I carried this work further. This chapter ends with a brief discussion of the plausibility of the role of white holes as generators of high-energy particles.

## WHITE HOLES AS PARTICLE ACCELERATORS

Astronomers encounter several instances of sources of high-energy particles or radiation in the cosmos. The highest-energy particles in the cosmic rays that bombard the Earth's atmosphere are protons, whose energy per particle is as high as $10^{20}$ electron-volts. To put this number in a proper perspective, we have to remember that this is about a *hundred billion* times the energy of the proton at rest. The large increase in energy arises because the proton moves with a speed very close to the speed of light. What mechanism caused the proton to move with such high speed? So far, the conventional sources known to astronomers – such as an exploding supernova (Chapter 4) or a rapidly rotating pulsar (Chapter 6) – come nowhere near producing such highly energetic particles.

Examples of radiation of high-energy photons have become known with the progress of X-ray and gamma-ray astronomy. Of special interest to us here are the 'burst' sources, that is, sources pouring out X-rays or gamma rays over very short intervals, of the order of a second or less. Any astronomical object showing changes over

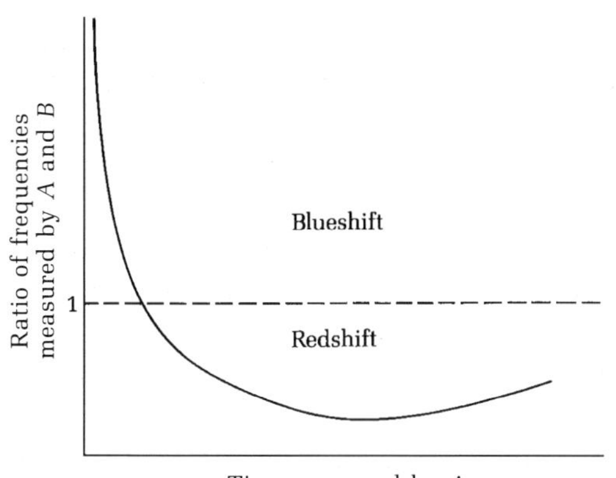


Figure 8-10. Instead of plotting the ratio of wavelengths measured by $A$ and $B$ against the white-hole size as in Figure 8.9, we now show how the ratio of frequencies measured by $A$ and $B$ changes with the time kept by $A$'s clock. The blueshifts last for a very short time in the beginning, which indicates that if a white hole is to generate high-energy particles and radiation, it will do so in a short time, in a burst. (Figure not drawn to scale.)

such a short time scale must necessarily be compact. A general rule of thumb used by the astronomer in setting a limit to the linear size $L$ of a source is the simple formula

$$L < c\,T.$$

That is, $L$ cannot exceed the product of the speed of light $c$ and the time scale $T$. For example, a source of time scale $T = 1$ second will have a size not exceeding 1 *light-second* (which equals 300 000 kilometres).

What could be the central mechanism that, acting over a compact region of space, produces matter and radiation of such high energies? Among the various possibilities considered by astronomers, the white hole offers a tempting alternative.

In Figure 8-9, we saw how a white hole produces very large blueshifts in light waves going *outward* from its surface. The large-blueshift phase, however, lasts only while the size of the white hole is very small. In Figure 8-10, we note that this phase lasts a very short time, and therefore the radiation poured out by the white hole does have the characteristics of a *burst.* A blueshift can be high enough in the early stages to convert visual radiation to X-rays or even gamma rays.

The same effect can also lift the energies of particles emitted from the surface of a white hole to the high values seen in cosmic rays. As with radiation, this effect operates only in the early stages of the white-hole explosion.

White holes could exist in astronomical objects displaying signs of explosion. Quasars, radio galaxies, and the nuclei of Seyfert galaxies are likely sites for white holes (see Figures 8-11 and 8-12).

## CAN WHITE HOLES EXIST?

As I stated before, white holes seem to do pretty much all that is required of black holes in the way of cosmic energy machines; and that too in a more direct way. All evidence to date points to outward flows of matter and radiation from sites of high-energy sources, and white holes describe these in a natural way.

In spite of their attractive possibilities in high-energy astrophysics, white holes have not gained the acceptance they deserve. What is the reason for the apparent neglect of white holes by astronomers?

There are several reasons why white holes have been poorly received. One easily understood reason is that we do not know what causes a white hole to erupt. That its initial state is of a singular character is clear. But how did this state come about in the normal course of astronomical development? For a black hole, on the other hand, a recipe for formation exists – at least in the case of stellar black holes. In Chapter 7, we saw that such black holes form toward the end of the evolution of a massive star.

In 1965, Y. Ne'eman from Israel and I. D. Novikov from the (former) USSR independently suggested that white holes are 'delayed' events representing creation of matter – delayed with respect to the epoch when the entire Universe was created some 10 billion years ago (see Chapter 9). The singular origin of a white hole is therefore a 'lagging' event that has taken place comparatively recently.

Work by D. M. Eardley in the United States and by K. Lake and R. C. Roeder in Canada seems to considerably restrict this *lagging core* scenario for white holes. The difficulty is this. In the early stages, as the white hole explodes, its outer surface may encounter ambient matter. Relative to the white-hole surface, this matter moves inward, and even a small density of this matter can smother the ten-

Figure 8-11. A photograph of the galaxy NGC 5128, identified with the radio source Centaurus A. This radio source (in common with many similar radio sources) may have originated in an explosion. (Courtesy of Palomar/California Institute of Technology.)

dency of the white-hole surface to move outward. The result is a slowing down of the outward motion and the ultimate collapse of the object into a black hole. These investigations therefore concluded that white holes are highly unstable objects, except when they occur soon after the origin of the Universe. As such, they will not be of much use in explaining more recent astronomical phenomena.

However, it seems premature to pass judgement on white holes on the basis of these objections alone. As we shall see in Chapter 11, it is now possible to provide an answer to the question of how a white hole originates. This alternative does not suffer from the 'smothering' problem of the lagging core scenario.

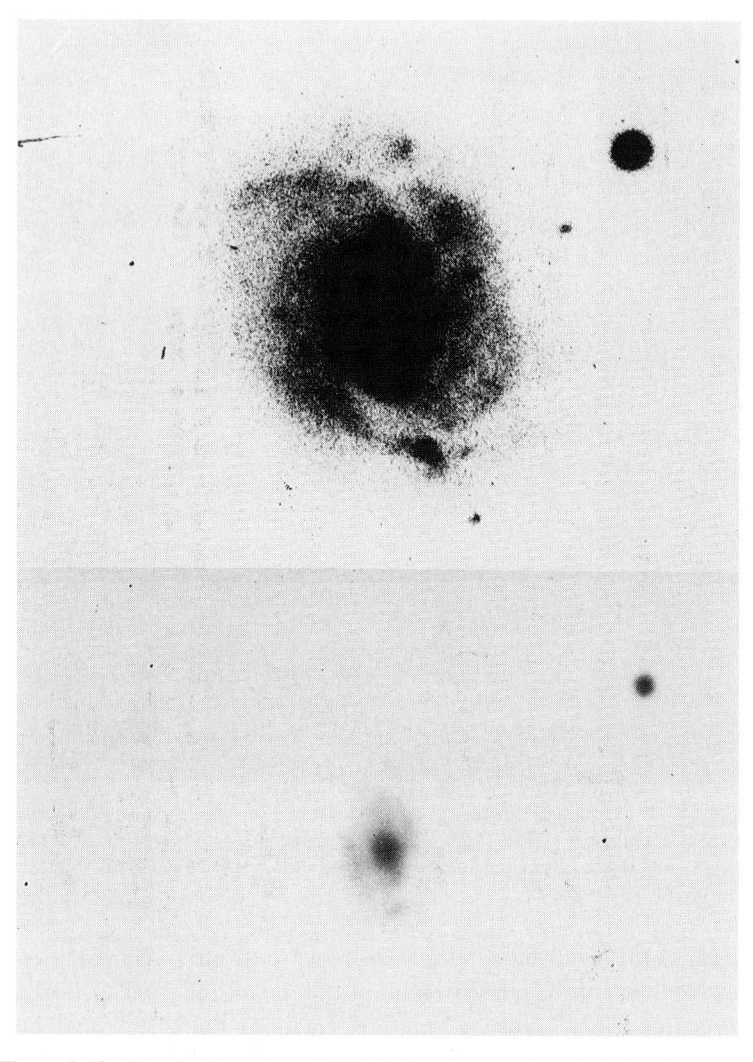

Figure 8-12. The Seyfert galaxy NGC 1068, whose nucleus is believed to show signs of explosion. The bottom half of the picture is underexposed to show only the nuclear region of the galaxy. (Courtesy of the Indian Institute of Astrophysics.)

In spite of these questions regarding white holes of a limited size, astronomers have, paradoxically, given an uncritical acceptance to the hypothesis of the biggest white hole of them all: the Universe! We discuss this remarkable hypothesis in the next chapter.

# 9
# *The big bang*

## THE STATIC UNIVERSE IN NEWTON'S THEORY

Back in the 1690s, Isaac Newton attempted an ambitious application of his law of gravitation. Newton wanted to describe, with the help of his theory of gravity, the largest physical system that can be imagined – the Universe. How did Newton fare in this attempt?

In a letter to Richard Bentley dated 10 December 1692, Newton expressed his difficulties in the following words:

> It seems to me, that if the matter of our Sun and Planets and all ye matter in the Universe was evenly scattered throughout all the heavens, and every particle had an innate gravity towards all the rest and the whole space throughout which this matter was scattered was but finite: the matter on ye outside of this space would by its gravity tend towards all ye matter on the inside and by consequence fall down to ye middle of the whole space and there compose one great spherical mass. But if the matter was evenly diffused through an infinite space, it would never convene into one mass.

Figure 9-1, which illustrates a finite and uniform distribution of matter in the form of a sphere initially at rest, helps explain Newton's difficulty. Will the sphere stay at rest forever? The matter in the sphere has its own force of gravity, which tends to pull the different parts of the sphere toward one another, with the result that the sphere as a whole contracts. We have encountered this force of self-gravity in stars (Chapter 4) and in the phenomenon of black-hole formation (Chapter 7). In the case of stars, the internal pressures oppose self-gravity and maintain the stars in a static shape. In the formation of a black hole, these pressures are negligible compared to

nificance to the position of the Earth, and from the Earth, observers had a unique perspective when looking north, south, east, or west. The Greek Universe was neither isotropic nor homogeneous.

Even with these simplifying assumptions about the large-scale structure of Universe, the *quantitative* details were still lacking in Einstein's model. To determine these details, Einstein needed his theory of gravitation – the general theory of relativity.

In Chapter 5, we saw how the geometry of spacetime is different from Euclid's geometry in the neighbourhood of a massive object like the Sun. It is the main feature of general relativity that any distribution of matter (and energy) should affect the geometry of spacetime around it. Einstein therefore expected that the distribution of matter (in the form of stars, galaxies, etc.) should determine the geometry of the large-scale structure of the Universe. But here he encountered one major difficulty.

The equations of general relativity permitted models of the Universe that were homogeneous and isotropic but *not static*. This difficulty is in fact no different from that which troubled Newton two centuries earlier: how can matter remain stationary in spite of its self-gravity? The quotation at the beginning of this chapter expresses Newton's difficulty within the framework of his theory of gravity.

To counter the self-gravity of the Universe, Einstein invented a new force of *repulsion*, known as the $\lambda$-force. According to Einstein, this force of repulsion increases in direct proportion to the distance between any two chunks of matter. The universal constant $\lambda$ determines the strength of this force of repulsion.

According to Einstein, the matter in the Universe is held in equilibrium under two opposing forces – the force of attraction of gravity and the $\lambda$-force of repulsion. Einstein found that such a Universe could be static, provided it was finite but unbounded.

Can an object be finite but without boundaries? Yes! For example, the surface of a sphere has a finite area, but where is its boundary? An even simpler example is that of a circle: its circumference is finite, but it has no beginning or end! The circle and the sphere are examples of one- and two-dimensional spaces. Can we similarly imagine a space of *three* dimensions that has a finite *volume* but no boundaries? Although our sense of perception does

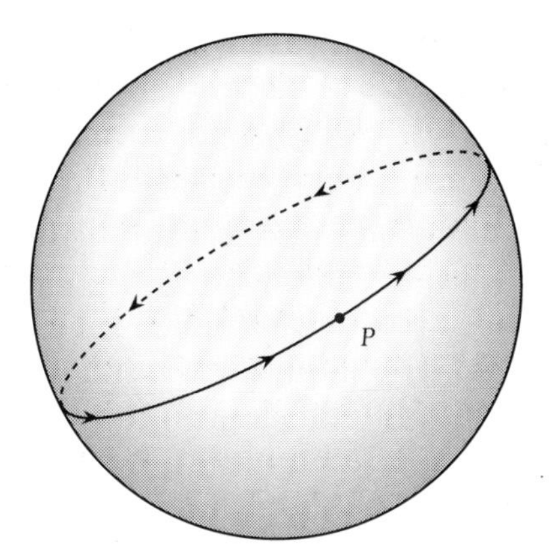


Figure 9-2. A two-dimensional example of how a light ray in the Einstein Universe would make a complete round of the Universe and return to the starting point. A great circle represents a 'straight line' in the non-Euclidean geometry on the surface of a sphere. A light ray constrained to move in the *two*-dimensional space of the surface of a sphere would travel along such a track. Thus, it would start from $P$ and end at $P$. We have to imagine a similar situation in the *three*-dimensional space of the Einstein Universe.

not help us to form an intuitive picture, we can rely on mathematics. A mathematician would tell us that the space we are trying to imagine is the three-dimensional boundary of a four-dimensional *hypersphere*. Just as the circle and the sphere are made of one- and two-dimensional distributions of points at the same distance from a fixed point in space known as the centre, so the hypersphere is made of a three-dimensional distribution of points, all of which are located at the same distance from the centre. If this distance, known as the radius of the hypersphere, is $R$, the three-dimensional space of the hypersphere has a volume

$$V = 2\pi^2 R^3.$$

This space has no boundary! Just as moving on the surface of a sphere – say, along a meridian (or any great circle) – we eventually come back to our starting place, so we would return to our starting point if we go 'straight' in the Einstein Universe (see Figure 9-2). How far would we have to move to make one circuit of the Einstein Universe? The answer is, the distance $2\pi R$.

Einstein's equations determine $R$ in terms of $\lambda$. The relation is a simple one:

$$R = \frac{c}{\sqrt{3\lambda}}.$$

The symbol $c$ here denotes the speed of light, which we can measure. If we know $\lambda$, we know $R$. How do we know $\lambda$?

Einstein's equations give another relation – between $\lambda$ and the average density of matter $\rho$ in the Universe. The relation is

$$\lambda = \frac{4\pi G\rho}{3}.$$

So if we know $\rho$, we know $\lambda$, and then we know $R$. Astronomers tell us that the average density of *visible* matter in the Universe is about $3 \times 10^{-31}$ grams (that is, three ten thousand billion billion billionths of a gram) per cubic centimetre. This is the density of matter in the form of galaxies, quasars, radio sources, etc. – objects whose existence is established by direct observations. If we also include dark matter, this estimate may well go up (see the next chapter).

Taking $\rho = 3 \times 10^{-31}$ grams per cubic centimetre, we get $R \cong 6 \times 10^{28}$ cm. Thus, one round of the Einstein Universe is approximately 350 billion light-years long. That is, light would take 350 billion years to make one round of the Universe! Einstein's original estimate, based on the information then available, was as low as about 10 million years.

Einstein's Universe was the beginning of modern cosmology. It pointed the way to further discussion of the large-scale structure of the Universe *within the framework of physics.*

## THE UNIVERSE IS NOT STATIC

The Einstein Universe was not, however, destined to remain for long an acceptable model of the actual Universe. In fact, within twelve years of its birth, astronomers began to discover evidence that contradicted its basic premise that the Universe is static.

In 1929, Edwin Hubble, an astronomer at the Mount Wilson Observatory (now part of the Mount Wilson and Las Campanas Observatories) near Pasadena, California, published a paper entitled 'A relation between distance and radial velocity among extragalactic

Figure 9-3. The 100-inch telescope at Mount Wilson, used by Hubble for his investigations of galaxies. (Courtesy of The Observatories of the Carnegie Institution of Washington.)

nebulae' in *The Proceedings of the National Academy of Sciences of the United States*. This paper presented a remarkable result that was discovered after several years of observations with the newly established 100-inch telescope at Mount Wilson (see Figure 9-3). The observations were based on a systematic study of the spectra of light from galaxies situated well beyond our own Galaxy.

These spectra revealed the redshift effect that we encountered earlier in a different context (Chapter 8). If we look for the familiar lines in the spectrum of a distant galaxy, we find them not at the wavelengths normally associated with these lines in a terrestrial laboratory but at longer wavelengths. For example, the H and K lines of calcium are expected to have wavelengths at 3933 Å and 3968 Å, respectively. (Å stands for the wavelength unit ångström, which is a hundred-millionth of a centimetre.) In the spectrum of a galaxy in the Hydra cluster, Hubble and his colleague Milton Humason found these lines at wavelengths 4537 Å and 4578 Å. Thus, from our earlier definition of redshift (see Chapter 8) as the fractional increase in the wavelength, this galaxy has a redshift

$z = 0.15$. In fact such redshifts had been spotted by other astronomers before Hubble, notably by V. M. Slipher between 1912 to 1925.

Hubble, however, found another remarkable property of these galaxies. He found that the fainter the galaxy, the larger is its redshift. Now, if we make the assumption (as Hubble did) that the galaxies around us have more or less the same luminosity, then the faintness of a galaxy is an indicator of its distance. The farther the galaxy is from us, the fainter it would look to us.

When Hubble took account of this relationship between faintness and distance, he was able to arrive at a rough method of measuring the distance of nearby galaxies. He then plotted the redshift of a typical galaxy against its distance from us. Figure 9-4 shows Hubble's graph of redshift $z$ against distance $D$. Hubble's observational points lay close to the straight line shown in the figure, a result that led him to predict a simple law relating $z$ to $D$:

$$cz = HD.$$

This law, now known as Hubble's law, can be interpreted as follows. If we assume that the redshift of a galaxy is due to the Doppler effect (as Hubble did), then $cz$ gives a measure of the speed of recession of the galaxy. Hubble's law then tells us that the galaxy is moving away from us with a speed that increases in proportion to its distance from us. The constant $H$, called Hubble's constant, tells us how fast a galaxy is moving at a given distance.

What is the value of $H$? Notice that $1/H$ has units of time. Hubble got approximately 1.9 billion years for $1/H$. His estimates for distance were, however, grossly understated. Thus, the modern values of $1/H$ are five to ten times bigger than what Hubble found. In 1994, the Hubble Space Telescope in a major finding announced a value of $1/H$ in the range 10–15 billion years.

Hubble's law, therefore, presents a picture of the Universe far different from the *static* Universe of Einstein. The Universe seems to be *dynamic*, with galaxies moving away from us, as if we are in a highly unpopular part of the Universe. Does this mean that we are back to some revised version of the old Greek cosmology that accorded a unique status to us on the Earth as observers of the cosmos?

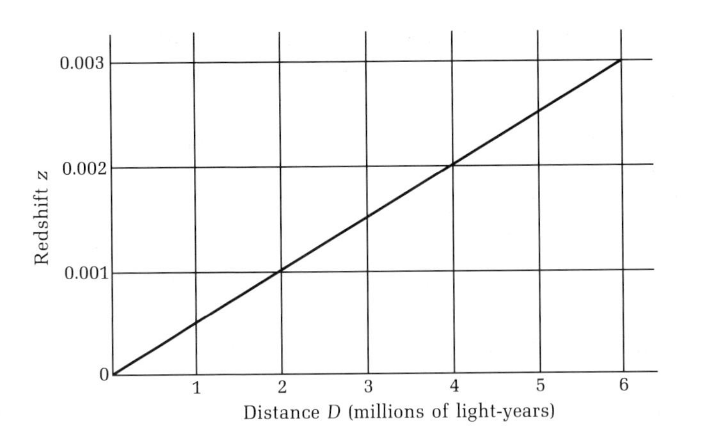


Figure 9-4. The straight line graph calculated by Hubble from his original measurements of the redshift $z$ and the distance $D$ of galaxies. Later it turned out that Hubble had grossly underestimated the distances of galaxies. The present estimates (which are still subject to uncertainties) make these galaxies five to ten times *farther away* than Hubble's original estimates.

## THE BIG BANG

Hubble's law does not, however, take us back to the pre-Copernican days. Although the redshift of a galaxy appears to increase with its distance from us, this does not necessarily place us in any unique position. In fact, rather the opposite is the case. If we imagine ourselves to be observing the Universe from another galaxy, we would find that the same recession phenomenon holds true with respect to our new vantage point. In other words, all galaxies would serve equally well as observation posts for the Hubble effect – our Galaxy does not enjoy any special status.

A common way of describing the recession of galaxies is to say that *the Universe is expanding*. The space in which the galaxies are embedded is expanding, so that the separation between any two galaxies is increasing. Figure 9-5 illustrates this expansion effect with two examples. (a) If we blow up a spotted balloon, the spots appear to move away from one another; yet there is no particular spot that can claim a special status. (b) A cubical grid of metal wires expands when heated, so that the lattice points of the grid move farther apart from one another. A third example is a special

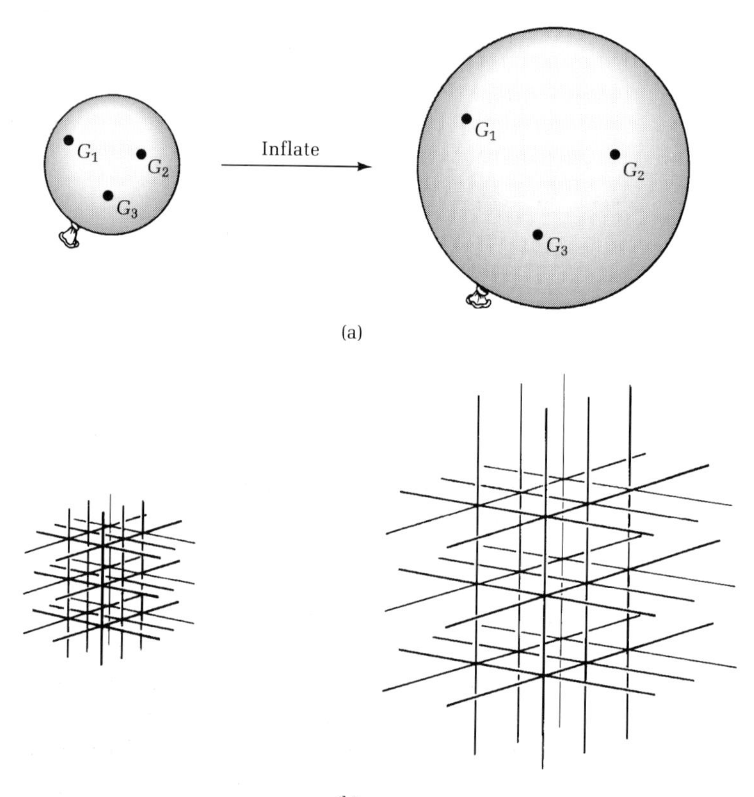


Figure 9-5. Two examples of expansion analogous to the expanding Universe. (a) As we inflate a balloon, the dots $G_1$, $G_2$, and $G_3$ on its surface move away from each other. No particular dot can claim a unique position on the surface of the expanding sphere. (b) A grid of metal wires expands when heated. The lattice points all move away from each other. Again, no one point can claim a privileged position within the grid. (From *The Physics–Astronomy Frontier* by F. Hoyle and J. V. Narlikar. Copyright © 1980. W. H. Freeman and Company.)

kind of plastic strip that expands to three times its size when heated in an oven. Any figure drawn on the strip will also expand. In Figure 9-6, we see such a strip with a few dots on it, before and after heating. The dots have moved away from one another as the galaxies do in the expanding Universe.

Once we accept the fact that the galaxies are receding from each

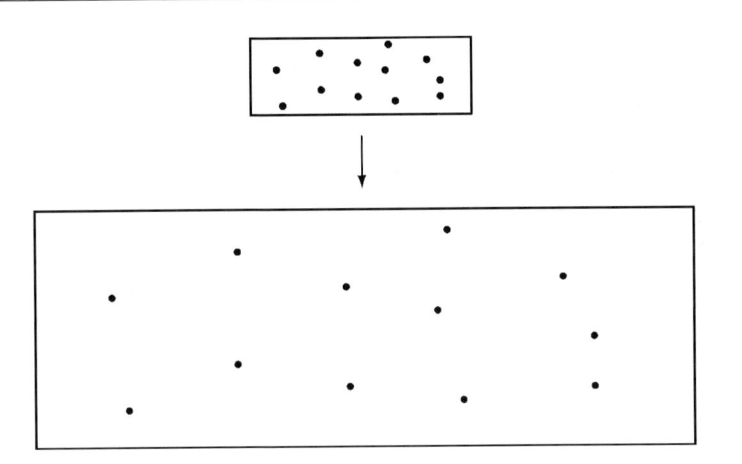

Figure 9-6. The dots on the plastic strip move apart from one another as the strip expands to three times its size (both in length and width).

other, the physicist will naturally ask what the cause of this expansion is. Has the expansion been going on in the past? Will it continue in the future? Questions like these can be answered with the help of mathematical models of the expanding Universe, just as Einstein constructed a model of the static Universe. Here again Einstein's general relativity comes to the help of the theoretician.

In 1922, seven years before Hubble published his results, the Russian physicist Alexander Friedmann had constructed such models. Friedmann used Einstein's assumptions about the homogeneity and isotropy of the Universe, but he dropped the assumption of a *static* Universe. The simplest Friedmann models do not require the $\lambda$-term introduced by Einstein for his static model; they are based on Einstein's old equations of 1915.

If gravity is the only force involved in these cosmological models, and if gravity has a tendency to contract a system of masses rather than to expand it, how was Friedmann able to construct models of an *expanding* Universe? The answer to this question can be understood with the help of a simple example. Recall our ball thrower in Chapter 3. The ball thrower sends a ball *upward* in the vertical direction. For a time the ball travels upward, although the force of Earth's gravity tends to pull it *down*. The reason the ball can go up at all is that it has been thrown with an *initial* upward velocity. It will continue to go up until its store of kinetic energy is completely exhausted.

In the same way, the Universe tends to expand in spite of gravity because in the *initial* stages it was given large kinetic energy as happens, for example, in an outward explosion. Imagine a gigantic cosmic explosion in which the components of the Universe were thrown apart. What we see today is the debris of this cosmic explosion, commonly known as the *big bang*. The Friedmann Universe was *created* with a bang.

Had there been no gravity, the speeds of recession of the galaxies would have continued unabated. However, the gravity of the Universe shows its effect by *slowing down* these recession speeds, just as the Earth's gravity slows down the ball's upward speed. We can follow this ball example further to determine the future behaviour of the Universe.

We saw in Chapter 3 that the ball falls back to Earth unless it is thrown with sufficient speed. The critical speed that determines whether the ball will fall back or not is the *escape speed*. If the escape speed is exceeded, the ball never returns to Earth; it keeps going away from Earth forever.

Similarly, there is a critical feature in the big bang that determines whether the Universe will keep on expanding forever, albeit with decreasing speed, or whether it will slow to a halt and fall back upon itself. As expected from the geometrical nature of Einstein's theory, this dynamical behaviour of the Universe turns out to be related to its geometry.

## IS THE UNIVERSE OPEN OR CLOSED?

While describing the static Einstein Universe we noted that its space is finite in volume but unbounded. A ray of light going in any direction in this Universe makes a complete round of space and returns to the starting point. A Universe with these properties is said to be *closed*. The curvature of space in such a Universe is positive (see Chapter 5).

The space in the expanding Friedmann Universe can have positive, zero, or negative curvature. We denote these three possibilities by $A, B$, and $C$, respectively. The Friedmann model Universe of type $A$ is closed, as is the Einstein Universe. The models of types $B$ and $C$ are *open*. They are *infinite and unbounded*. All three types, however, satisfy the condition of homogeneity and isotropy at any epoch. That is, at any epoch we can locate observers all over the Universe who will report the same large-scale view of the cosmos. A typical observer will also find that the Universe looks the same in all directions.

These large-scale views describe the structure of the Universe on the scale of, say, a few million light-years or so. On a smaller scale, we do notice inhomogeneities, such as the shape of our Galaxy, our own off-centre location in it, etc. These 'local irregularities' are not important to discussions of the dynamics of the Universe on the large scale. We will, however, consider these complications in the following chapter. For the time being we proceed on the basis of our simple assumptions.

How the differences of geometry affect the expansion of the Universe is shown in Figure 9-7, where $S$ is the characteristic *scale factor* of the expanding Universe. The relative linear separation between two typical galaxies changes in proportion to $S$, so that an increase of $S$ with time signifies the expansion of the Universe.

In Figure 9-7, all three curves for models $A$, $B$, and $C$ start off with $S = 0$ and have $S$ subsequently increasing with time. In the case of $A$, however, the expansion slows to a halt and gives way to contraction. Thus, the closed model of the Universe has a *contraction phase* following the expansion phase. The open models continue to expand forever. Compare this behaviour of the Universe with the flight of the ball we discussed earlier.

All three curves show the expected slowing of expansion due to gravity. The epoch at which $S$ was zero is known as the *big-bang epoch*, the instant of creation when all the matter in the Universe was compressed in zero volume. This is the singular instant that marks the origin of the Universe. We may start the cosmic clock at this instant, so that in Figure 9-7 the time axis reads $t = 0$ at $S = 0$.

The singular epoch at $t = 0$ and the subsequent expansion of the Universe bear a great similarity to the white-hole theory of Chapter 8. At the singularity, any finite chunk of the Universe is like a white hole erupting from a point in space. This similarity led to

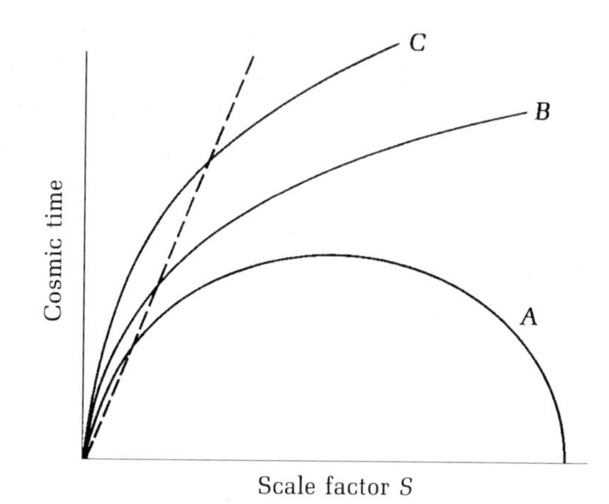


Figure 9-7. The three types of cosmological model. Model $A$ describes a closed Universe, whose expansion slows down and gives way to contraction. $B$ and $C$ are models of an open Universe, which goes on expanding forever. The dashed lines represent the expansion of the Universe in the absence of gravity.

the suggestion that white holes may be 'delayed bangs' – eruptions that took place in space later than the moment $t = 0$.

The dashed straight line of Figure 9-7 shows how $S$ would have behaved had there been no gravity. In fact, there is a Friedmann model that shows this rate of expansion, but it is completely empty. The greater the density of matter in the Universe, the greater is the effect of gravity on slowing the expansion. We therefore expect the closed models to be *denser* than the open models. There is in fact a critical density that separates the closed models from the open ones. At the present epoch, this density, sometimes known as the *closure density*, is given by

$$\rho_c = \frac{3H^2}{8\pi G}.$$

If the actual density in the Universe *exceeds* $\rho_c$, the Universe is closed; if it equals $\rho_c$ or is less than $\rho_c$, the Universe is open. The three models $A, B$, and $C$ of Figure 9-7 have $\rho > \rho_c$, $\rho = \rho_c$, and $\rho < \rho_c$.

The question of whether the Universe is open or closed can in principle be decided if we have reasonably accurate measures of the density $\rho$ of matter in the Universe and the Hubble constant, $H$. Unfortunately, neither $\rho$ nor $H$ are known accurately. We earlier gave a value of $\rho \cong 3 \times 10^{-31}$ grams per cubic centimetre for the visible matter in the Universe together with at least a comparable component of dark matter. We also remind ourselves that on the basis of the present data, the range within which $1/H$ is expected to lie extends from 10 billion to 15 billion years.

Taking the upper end of this range, we calculate the density to be $\rho_c \approx 10^{-29}$ g cm$^{-3}$. If we had taken the lower end of the range, we would have arrived at 2.25 times this value of $\rho_c$. Notice that in either case $\rho < \rho_c$, and so we have a prima facie case for the open Universe of type $C$.

However, the issue of whether the Universe is open or closed is not as simple as that! The proponents of the closed Universe argue that the estimate of $\rho$ given above is only a lower estimate. If there is considerably more dark matter in the Universe, the estimate for $\rho$ may be higher than the above value. If it becomes high enough to exceed $\rho_c$, the Universe may be closed.

Another way to decide which of the models $A, B$, or $C$ is the correct one is to measure the rate at which the expansion of the Universe is slowing down. The rate of slowing down is higher for $A$-type models and lower for $C$-type models than it is for $B$-type models. However, this is a difficult measurement to make, and it has not yet been possible to arrive at unambiguous conclusions about the slowing down of the rate of expansion.

## WAS THERE A BIG BANG?

Cosmologists are often asked whether there is any direct astronomical evidence for the big bang.

When astronomers survey the distant parts of the Universe, they do not see the Universe as it is *now* but as it was in the remote past. This happens because astronomical observations depend on light, which has a finite velocity. Light that carries information from a distant source takes some time to get to the astronomer. For example, light from a galaxy a billion light-years away from here takes a billion years to reach here. Consequently, the astronomer sees the galaxy as it was a billion years ago.

This delay due to the finite speed of light is hardly noticeable in our day-to-day observations on the Earth because the distances are small. However, people making international telephone calls routed via satellite do notice a momentary delay in conversational responses, which are carried by radio waves with the speed of light.

Because the Universe has been expanding since its origin at $t = 0$, the scale factor $S$ was smaller in the past than it is now. Is there any direct way we can compare the scale factor a billion years ago with the scale factor at present? Yes! A simple result from Einstein's general theory of relativity enables us to do this. Suppose we measure the redshift of a galaxy at the distance of a billion light-years and call it $z$. Then the ratio of the present scale factor to the scale factor a billion years ago is just the quantity $(1+z)$ (see Figure 9-8).

Thus, the larger the redshift of the observed object, the smaller is the scale factor of the Universe when the light ray left that object. The big-bang epoch corresponds to the epoch of infinite redshift.

The largest redshift of a quasar measured to date is $z = 4.897$. There is currently some controversy as to how far away quasars are. But even assuming that they are as distant as they appear to be, the farthest known object does not take us all that far back into the past!

There is, however, indirect evidence that takes us much farther back in the history of the Universe. We end this chapter with a brief discussion of this important evidence.

We have so far concerned ourselves with only the geometrical aspects of the Universe. Cosmology goes beyond this; it is also concerned with the physical aspects of the Universe. In the 1940s, George Gamow first discussed the physical state of the Universe just a few moments after the big bang.

The big bang implies violent activity, and physicists have wondered what form this activity took. Such speculations take us to the most elementary states of matter – perhaps even to *quarks*, which may have existed before electrons, protons, and other elementary particles were formed. Gamow's discussion, however, began at the stage after these particles were formed. He argued that in primordial times one indicator of the violent activity in the Universe was its temperature. The temperature was enormously high to begin with and declined steadily as the Universe expanded, much as a ball of hot gas cools down as it expands.

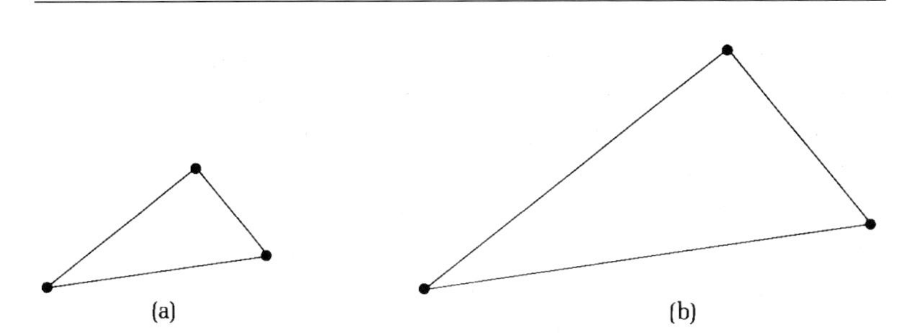


Figure 9-8. In (a) we see a triangle formed by three galaxies. In (b) the triangle has expanded to twice its original size, so that each galaxy is twice as far away from the other two galaxies. This is the result of the expansion of the Universe. If light from any of these galaxies left at the epoch of (a) and was received at the epoch of (b), this light would show a redshift $z = 1$. (This result is a special case of the general result described in the text.)

In the early stages, elementary particles existed in a heat bath of extremely high temperature, of the order of 10 billion kelvin, just 1 second after the big bang. This early stage had another remarkable property: it featured the dominance of radiation over matter, in contrast to the present state of the Universe, which is dominated by matter. Figure 9-9 shows how the transition from the early radiation-dominated era to the later matter-dominated era occurred in the course of the expansion of the Universe.

Recall that we have encountered temperatures as high as millions to billions of kelvin inside stars (Chapter 4). We noted then that at such temperatures stars run highly efficient thermonuclear fusion reactors that successively make nuclei ranging from helium, carbon, oxygen, etc., all the way to iron. Now, in the early history of the big-bang Universe, during the period from 1 second to 200 seconds, the temperature passed through a similar range. In this context George Gamow therefore made the remarkable suggestion that the various elements that we see around us were in fact cooked in the hot early Universe.

Gamow's hunch was backed by calculations. However, in retrospect we find that the claim was only partially correct. The conditions in the early Universe were suitable for making deuterium and helium and a few other light nuclei in very tiny amounts; but beyond that (from carbon onwards) they were not. For making those heavier elements one indeed needs stars.

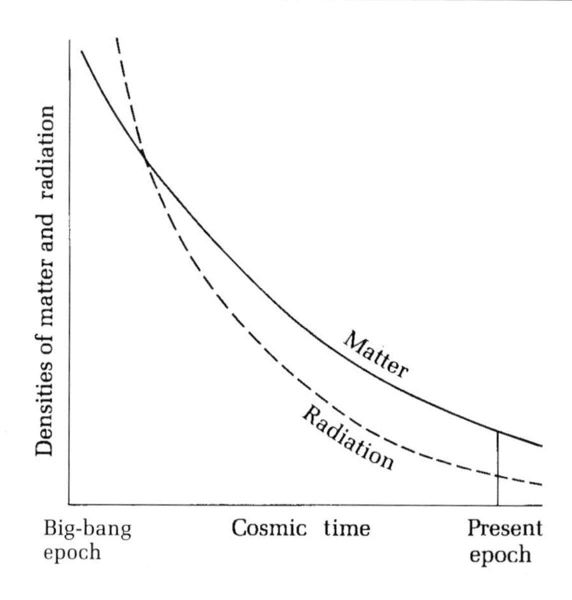


Figure 9-9. A schematic diagram showing how the density of matter (*solid curve*) and radiation (*dashed curve*) decrease as the Universe expands. In the early states, the Universe was dominated by radiation; the present epoch is dominated by matter. This is because as the Universe expands, the decline in the density of radiation is more rapid than the decline in the density of matter. (Figure not drawn to scale.)

However, the Gamow version of nucleosynthesis seems to fill the gap left by the stellar version. For stars are not able to make deuterium and the amount of helium they make is far less than the quantity of helium observed. Approximately one-quarter of all the mass of observed matter is in the form of helium, and it seems that the bulk of it must have come from the early hot Universe. There is likewise reasonable agreement between observations and the theoretical predictions of deuterium and some other light nuclei such as lithium, beryllium, and boron. Thus, Gamow had definitely scored a partial success with his hot Universe scenario. But he was even more successful with another of his predictions.

Gamow conjectured that the relic of the early hot radiation should now be seen in the form of cool radiation. How cool? On the basis of the information then available, Gamow's colleagues Ralph Alpher and Robert Herman predicted the present temperature of the

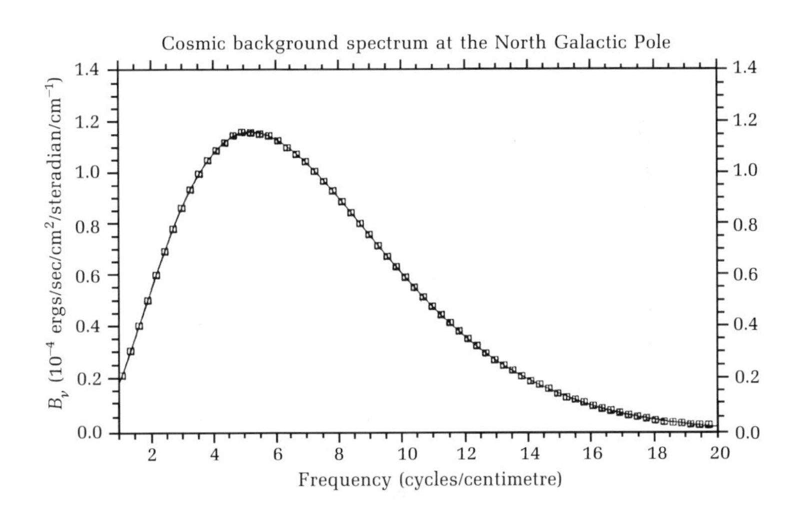


Figure 9-10. The spectrum of microwave background radiation as measured by COBE. The spectrum tells us how the energy of radiation is distributed over different wavelengths. The continuous curve passing through the 'error boxes' represents the theoretical expectation from a big-bang origin of the radiation. Such radiation is said to be of black-body type, that is, it has the energy distribution characteristic of radiation trapped within an enclosure for a long enough time. The temperature corresponding to the continuous curve is 2.73 kelvin on the absolute scale of temperature. (Courtesy of the COBE team and National Aeronautics and Space Administration, USA.)

radiation to be about 5 kelvin*. At this temperature, the radiation would exist predominantly in the form of *microwaves*.

The first important indications of the existence of such radiation came in 1965, when Arno Penzias and Robert Wilson at the Bell Telephone Laboratories in Holmdale, New Jersey, accidentally found a radiation background at 7 cm wavelength, with a temperature of about 3.5 on the absolute scale. The possible cosmological significance of this discovery prompted several groups of astronomers to measure the radiation background at other wavelengths. Today measurements exist at wavelengths ranging from 70 cm on the long-wavelength end to wavelengths shorter than

* This important prediction of relic radiation was made by Alpher and Herman in a paper in *Nature* (vol. 162, p. 774, 1948). Gamow himself had made a guess of a temperature of 7 kelvin in a different publication in 1953.

1 mm. The latter measurements became possible only with recent advances in space technology.

The Cosmic Background Explorer (COBE) satellite launched in 1989 came up with the most comprehensive data on the spectrum of radiation. The characteristics of this radiation are those expected from the radiation left over from the big bang (see Figure 9-10). The temperature is not too far off the value roughly guessed by Gamow and his colleagues. The radiation appears as a background – that is, it seems to come equally from all directions in the Universe, thus suggesting its cosmic origin.

Results from the COBE satellite have therefore given a boost to the big-bang hypothesis. Thus to the question heading this section most astronomers will reply in the affirmative. Some even go so far as to say that the 'ultimate problem' of the origin of the Universe has been solved. This is perhaps an overstatement, for as we shall see in the next chapter the Universe has still quite a few mysteries up its sleeves.

# 10

# *The Universe: from simplicity to complexity*

## THE PROBLEMS OF LARGE-SCALE STRUCTURE

More than seven decades have elapsed since Friedmann proposed his mathematical models that describe the expanding Universe. As we saw in Chapter 9, these models lead to the conclusion that the Universe was created some 10–15 billion years ago in a big explosion (the so-called *big bang*) after which it has been expanding but more and more slowly because of brakes applied by gravity. This model also tells us that the Universe was very hot to begin with, and dominated by radiation, but with expansion it has cooled down and the temperature of the radiation background today is 2.73 kelvin (see Figure 9-10) as measured by the COBE satellite and other ground-based detectors. And one other set of relics of the hot era, namely the light nuclei like deuterium, helium, etc., are found in the right amount all over the Universe. Thus, we concluded the last chapter with a fair degree of confidence in the big-bang scenario.

However, over the last quarter of a century astronomical observations have become more sophisticated and the views of the large-scale structure of the Universe they present go well beyond the simplified assumptions of a 'homogeneous and isotropic Universe'. We shall see, for example, in Figure 10-1 how galaxies are distributed over the sky in depth. The dots in the figure represent galaxies and their distribution is clearly not smooth, as a homogeneous Universe would have us believe. We see filamentary structures separated by void regions (with practically no galaxies therein), much like the air

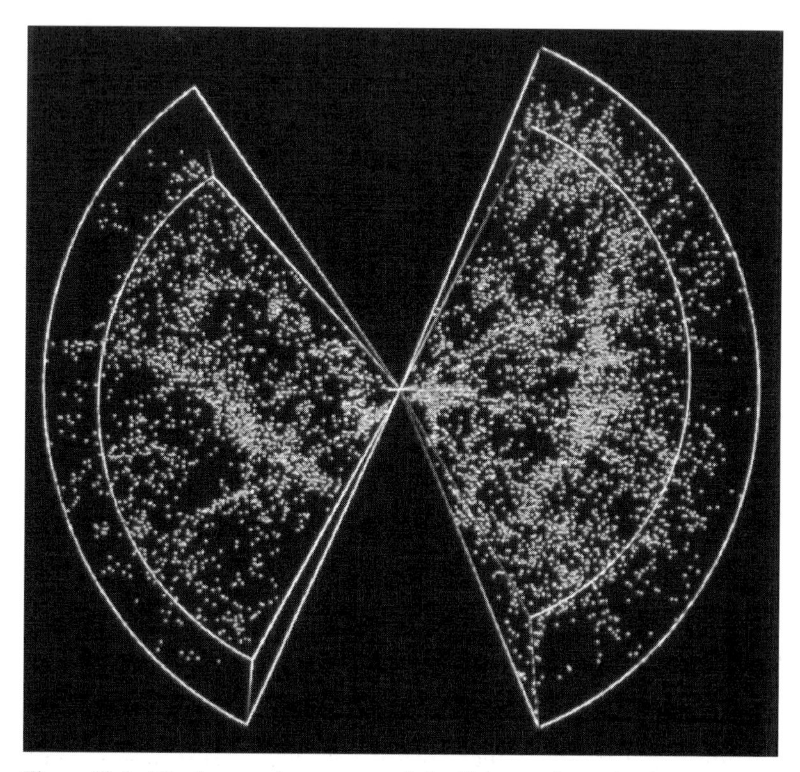

Figure 10-1. The large-scale structure of the Universe shown in the map of the nearby region towards the North and South Poles of our own Galaxy. The long sheet of galaxies in the northern portion is commonly known as the 'Great Wall'. (Photograph by courtesy of Margaret J. Geller, John P. Huchra, Luis A. N. da Costa and Emilio E. Falco, Smithsonian Astrophysical Observatory ©1994.)

bubbles in a sponge. The first question to ask, therefore, is, How did such inhomogeneous structure come about?

A second question faces us when we look at the velocities of these galaxies. If Hubble's law were right, we ought to see the velocity–distance law of Hubble prevail with systematic streamlined motions of all galaxies. Of course, small deviations from the Hubble motion are expected because of local gravitational effects. But astronomers were not prepared for what they found: systematic large-scale motions exceeding 500 kilometres per second in different directions, over and above the Hubble motion. So our second question is, How do we account for the origin and persistence of these motions?

Now we come to our third and final question related to the large-scale structure, but it is directed more at the microwave background. Unlike the distribution of galaxies, the radiation background is remarkably smooth. Of course, a smooth background is perfectly consistent with our simple assumptions of a homogeneous Universe. But when we look at it in juxtaposition to the patchiness of galaxy distribution, we are faced with the inevitable question, If matter and radiation both started off with a big bang and if both were interacting in the early stages, why has the former developed into a sponge-like structure while the latter has remained smooth?

## FOOTPRINTS ON THE SANDS OF TIME

In the big-bang scenario the three questions are believed to be interlinked. Take, for example, the interaction of matter with radiation. When the Universe was dense and hot this interaction was intense, with particles of radiation, the photons (see Chapter 7), frequently colliding with particles of matter, the electrons and vice versa. The calculations show, however, that when the Universe cooled down to about 3000 kelvin, the electrons slowed down sufficiently to be susceptible to the attractive electrostatic force of atomic nuclei, with the result that most of them became trapped inside atoms. The electrons had played the key role in the scattering of radiation, and with them out of the way, the radiation could move straight and unhindered by scattering. From then on matter and radiation each went along separate ways with no mutual interaction.

So the smooth radiation background we observe today has remained smooth from that earlier epoch of the trapping of electrons. Our question concerning the smoothness of radiation today therefore gets transferred to that earlier epoch. Just how smooth should the radiation have been at that epoch? This is where another discovery made by COBE in April 1992 becomes significant.

Figure 10-2 shows a computer colour picture of the fluctuation of temperature observed in the microwave background radiation by COBE. The image processing software assigns different shades of colour to different magnitudes of these fluctuations. But what is the overall order of magnitude of these fluctuations? Only a few parts in a million!

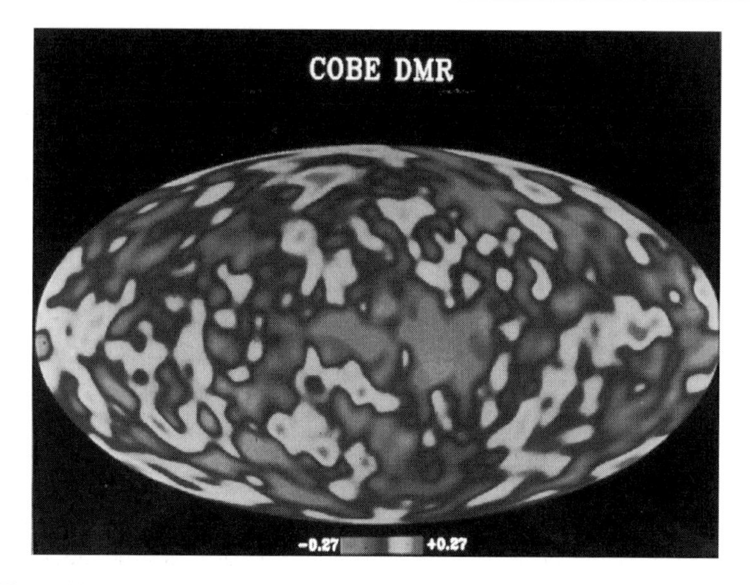


Figure 10-2. The tiny fluctuations of temperature of the microwave background observed by COBE. (Courtesy of the COBE team and National Aeronautics and Space Administration, USA.)

Perhaps an analogy will explain the significance of these tiny fluctuations. Imagine a beach with vast expanses of sand totally undisturbed and smooth. Observing such a beach, one will find it hard to convince anybody that the beach is frequently used. Why are there no footprints of humans or other animals? The sceptic may well ask this question.

Earlier studies – and there had been several – had failed to reveal any fluctuations of temperature in the microwave background down to the levels of a few parts in a hundred thousand. Thus, it was becoming hard to believe that the radiation background we see today had anything to do with the primordial seeds of fluctuation in matter density which might have somehow grown to the present panorama of galaxies. But the COBE discovery held out the hope to the theoreticians that a scenario like the following might possibly work.

Imagine the very early stages after the big-bang when both matter and radiation were in constant interaction. But both had very tiny seeds of fluctuations that were *comparable in magnitude*, like tiny ripples on an otherwise smooth surface of water. As the Universe expanded, its two components were evolving side by side, until the temperature dropped to the critical figure of about 3000 kelvin when the two ceased to interact. Thereafter, the radiation retained its imprints (now seen by COBE) while the matter fluctuations continued growing... till they became galaxies and groups and clusters of galaxies as we see them today.

## GRAND UNIFIED THEORIES AND INFLATION

The big-bang recipe for forming large-scale structures outlined above has two ingredients still missing. To appreciate the first of these let us look at the model Universe at a very early epoch. How early? Only a billion billion billion billionth part of a second ($10^{-36}$ second) after the big bang.

Human reactions cannot appreciate such a tiny measure of time. Even the best clocks so far provided by nature, atomic clocks in our laboratories or pulsars in space (Chapter 6), fall far short of measuring such a tiny period. How then can we make any sense of the above remark?

While experiments and observations set limits to human capabilities, theoretical aspirations can be unbounded. It is the theoreticians who have thought of a possible significance of this epoch in the history of the Universe. To them this epoch is a logical outcome of their search for unification of the various physical interactions.

In everyday life we encounter operations of the forces of electricity and magnetism. To a beginner, the needle of a magnetic compass always pointing north and an electric bulb lighting up at the flick of a switch appear to be two unconnected phenomena. Yet, the physicist knows that the forces of electricity and magnetism are interlinked, that they form parts of a single electromagnetic interaction. The appreciation of this unified picture first came about in the 1860s with the work of James Clerk Maxwell. And today physicists are in search of a similar unified picture that brings together the electromagnetic interaction and the strong and weak interactions (which correspond to forces that operate amongst subatomic particles at close range).

A partial success towards this goal was achieved in the 1970s when an 'electroweak theory' unifying the electromagnetic and weak interactions emerged from the works of Abdus Salam, Steven

Weinberg, and Sheldon Glashow, and was experimentally verified by the high-energy particle accelerators at CERN in Switzerland and Fermilab in the United States. The energy at which this unification takes place is about a hundred times the energy equivalent of the mass of a proton. Such energies, though very high, are within the production capability of these accelerators.

The next logical step now seeks a unification of the electroweak theory with the strong interaction which binds neutrons and protons into tight units, the nucleus of a typical atom (Chapter 4). A theory that will achieve this is called a grand unified theory (GUT).

While theoretical models of GUTs are available, their experimental verification seems well nigh impossible, since the energies of particles that would participate in a grand unified interaction are far higher – by a factor of a million million – than can be produced by the most advanced accelerators on Earth. Are GUTs therefore doomed to remain in the realm of unverifiable physical theories?

Almost; except for the possibility opened out by the big-bang model. If we extended our investigations far enough into the past and close enough to the big-bang epoch we should find particles moving with large enough energy to activate GUTs. And that 'close enough' epoch is the one mentioned in the first paragraph of this section. We will call it the *GUTs epoch*. In other words, very very early on, the Universe itself was a grand laboratory containing material particles moving at speeds high enough to test the workings of a grand unified theory.

The reader may wonder why we have kept quiet about gravity in this unification scheme. In fact Einstein himself, after the early successes of general relativity, had sought to unify gravity with the electromagnetic theory, but his attempts remained inconclusive. And today's theoretical physicists appreciate the rather unusual nature of gravity and are generally wary of including it in the GUT programme. However, the hope is that once the GUT is found, the next step to include gravity may be easier.

Nevertheless, as we have pointed out earlier (Chapter 5), gravity is all-pervasive and cannot be ignored. So, if the GUTs bring about important physical changes in the Universe $10^{-36}$ second after the big bang, those changes are bound to affect the geometry of spacetime, as required by general relativity.

D. Kazanas (1980), Alan Guth (1981), and Katsuhiko Sato (1981) independently came to this conclusion and discussed the impact these changes would produce on the expansion of the Universe. Their general deduction was that the Universe, while these changes are taking place, would expand far more rapidly than in the earlier Friedmann model. Guth gave this rapid expansion phase the apt name 'inflation'. Just as in an inflationary economy prices rise rapidly, so the Universe grows rapidly in size during the inflationary phase.

This phase, however, does not last long. The parts of space where the changes brought about by GUTs are over then slow down to the normal Friedmann rate. But in this brief interval those parts may have expanded by a tremendous factor. And this rapid expansion solves one of the outstanding problems of big-bang cosmology, called the *horizon problem*.

To appreciate the horizon problem let us digress a little and think of human societies in primitive times when communications were very limited. Thus, different cultures and civilizations separated by even a few hundred kilometres could be different in character. These local differences became blurred, and there was a trend towards uniformity as the modes of communications improved with frequent exchanges of ideas and information.

Our Universe must have passed through a similar 'uniformizing' process, in which communications by physical interactions tended to bring about a uniformity of physical conditions. However, as we have repeatedly emphasized, there is a speed limit to all interactions: they cannot travel with a speed exceeding that of light. This naturally limits the sizes of regions which could achieve uniformity. Thus, 1 second after the big bang, the typical distance over which one expects uniform conditions is 1 light-second, that is, the distance of 300 000 kilometres covered by light in 1 second. This distance is called the size of the 'particle horizon' at that epoch. Naturally the particle horizon grows in size as the Universe ages (see Figure 10-3).

However, if all the physical properties of the Universe, including its observed uniformity, were set at the GUTs epoch, the characteristic size over which conditions would be homogeneous is extremely small, not even a billion-billion-billionth part of a metre.

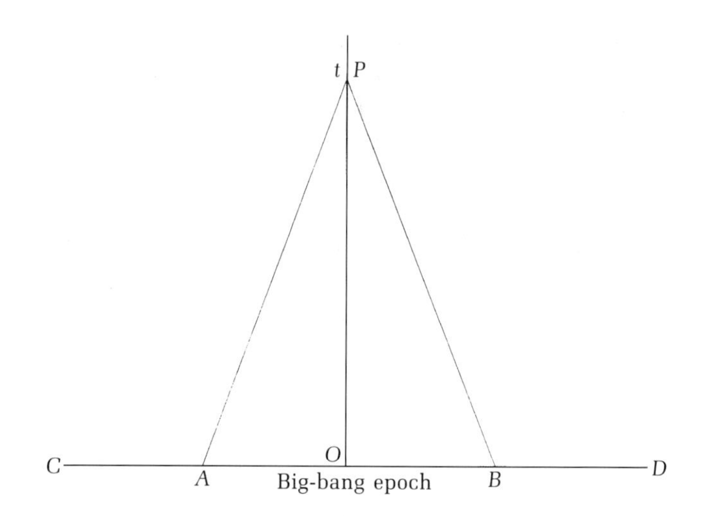


Figure 10-3. A spacetime diagram of the early Universe. $P$ is a typical observer at an epoch $t$. The two lines $PA$ and $PB$ representing paths of light rays hit the horizontal axis of the graph at $A$ and $B$. Since this axis denotes the epoch $t = 0$, this figure tells us that the light signals emitted by $A$ and $B$ will reach $P$ at time $t$. $P$ can be causally influenced by all points on the stretch $AB$ (like point $O$) but not beyond it (like points $C$ or $D$). If we rotate the diagram about the vertical axis we get a 'light cone' from $P$, all points within which affect $P$. The base of the cone is the particle horizon of $P$.

Although the Universe has since expanded, the corresponding size today would be less than a few centimetres! How, then, do we find the Universe to be homogeneous over 10 billion billion billion centimetres? This illustrates how severely the horizon limits the homogenization process.

Inflation offers a solution to this problem. It inflates the horizon size at the GUTs epoch by a very large factor that more than compensates the very small size we arrived at above.

It was because of this and other instances where the idea of inflation proved successful that it has become very popular with big-bang theorists. Further, it offers some promising pathways for the problem we started with, namely how to grow tiny primordial inhomogeneities in the density of matter to the gigantic sizes of galaxies and their clusters observed today.

Once you assume that the Universe passed through an inflationary phase around the GUTs epoch, you are also bound by one prediction of the theory, namely that the Universe today very closely conforms to curve $B$ of Figure 9-7 with its density equal to the closure density. We saw that the density of all luminous matter today corresponds to a few percent of the closure density. So there is an apparent discrepancy between what the inflationary scenario predicts and what is observed. How do theorists get round this problem?

To find the answer, we have to consider the next ingredient of the recipe for large-scale structure.

## DARK MATTER

In most countries there are two economies in operation. The first one is visible and legal, being based on *white money*, whose supply and demand are well documented and declared to the tax authorities. The second parallel economy is run on so-called *black money*, which bypasses the tax hounds. The extent of black money in circulation can be estimated, however, from its visible impact on the economy, such as construction activity, electioneering, advertising, movies, etc. The ratio of black to white money in circulation varies from nation to nation.

The reason for this apparent digression from cosmology is to highlight the similarity between black money and what the astronomers called *dark matter*. Dark matter, as the name implies, includes all forms of matter in the Universe that cannot be seen by astronomical methods, i.e., by any kind of telescope, optical, radio, or what have you.

We have already encountered one form of dark matter, namely the black hole. A black hole cannot be seen but its presence can be inferred from the dynamical activity it generates in its vicinity by its strong gravitational field. Indeed, gravity is the main (and in most cases the only) means of inferring the existence of dark matter, for as we saw earlier (Chapter 5), gravity is a universal property shared by *all* forms of matter. Black holes may be present in binary star systems and, on a more massive scale, in the nuclei of galaxies and quasars. Are there other forms and locations of dark matter?

In the 1970s astronomers began to examine the structure of individual galaxies by the technique of 21-centimetre wavelength waves.

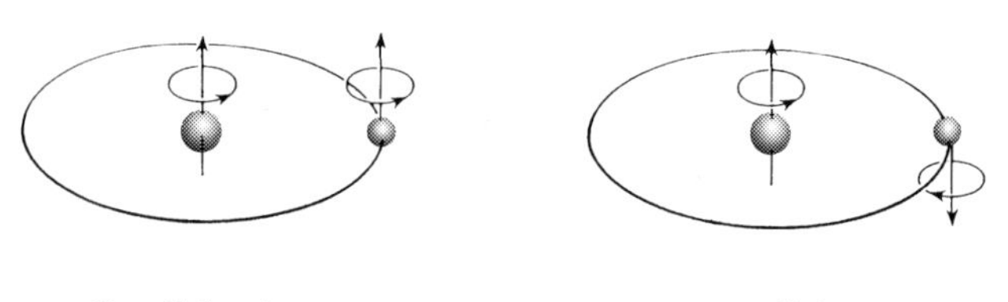


Figure 10-4. On the left we have the proton and the electron of the hydrogen atom spinning parallel to one another, while on the right their spins are anti-parallel. The latter combination has the lower energy. Radiation of 21 cm wavelength is emitted when the electron switches its spin from the former state to the latter state.

As shown in Figure 10-4, this particular radiation is characteristic of the hydrogen atom – the most common constituent of our Galaxy. This atom has one electron going round its nucleus, which contains only one proton. The atom can exist in two possible states. The state wherein both the particles are spinning in the same direction has higher energy than the state in which their spins are in opposite directions.

Nature has a tendency to induce physical systems to go for the states of lowest energy. The hydrogen atom, therefore, tends to go from the higher to lower energy state with its electron flipping over its direction of spin. The energy lost by the atom appears as a radio photon of wavelength 21 centimetres. Astronomers cash in on this result to use this radiation for detecting clouds of hydrogen in space. Moreover, by using the Doppler effect (Chapter 7) they can measure the speed with which a hydrogen cloud approaches us or recedes from us.

Measurements like these around galaxies, including our own enabled astronomers to measure the rotational speeds with which clouds of hydrogen go round galaxies. The reason why such clouds go round galaxies is the same as that which makes the planets go round the Sun, namely the force of gravity. By using Newton's and Kepler's laws, astronomers can infer the mass of the galaxy whose gravity is responsible for the rotational motion of the clouds. And these calculations led them to an unexpected result.

Let us first see what *was* expected. In Figure 10-5 we have hydrogen clouds going round in a galactic environment. In Figure

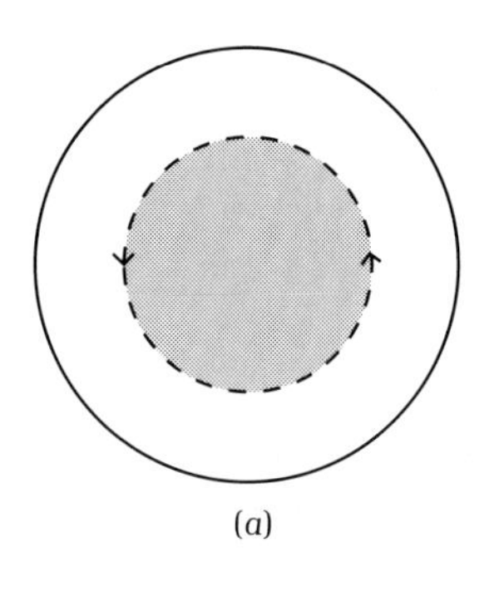


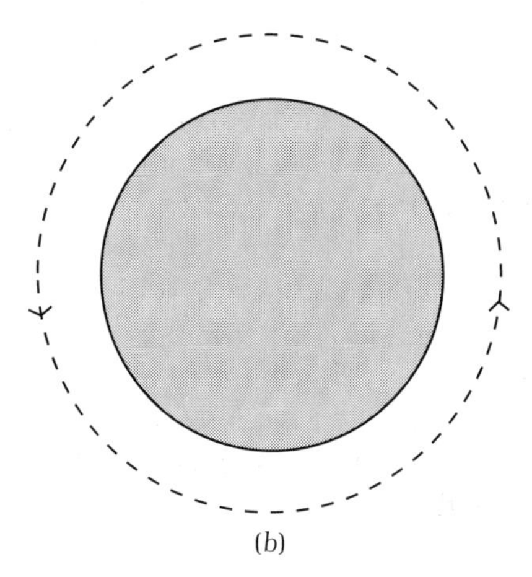


Figure 10-5. (a) Newton's law of gravitation tells us that for an interior circular orbit in a galaxy the attracting mass (shown shaded) is that interior to the sphere with the orbit as diameter. (b) For an exterior orbit the gravitating mass (shown shaded) is constant and is the mass of the galaxy.

10-5a the cloud is on a circular path concentric with and inside the galaxy (assumed spherical). For such a trajectory the effective attracting mass is that contained within the diameter of the trajectory. In Figure 10-5b the trajectory lies outside the galaxy whose entire mass attracts the cloud. Thus, clouds at different distances will have different rotational speeds.

Using this model, astronomers expected the rotational speeds to increase with distance from the centre of the galaxy for clouds moving *within* the galaxy and to decrease for clouds *outside* the galaxy. Figure 10-6a illustrates this behaviour by a curve.

Now comes the surprise. The curve of rotation speeds at different distances for a typical galaxy rises from the centre as expected, but then levels off, staying constant out to distances *well beyond the visible extent of the galaxy*. The real curves are shown for comparison in Figure 10-6b.

What are these curves trying to tell us? The first interpretation is that if all the matter is contained in the visible boundary of the galaxy, then the law of gravity, whether of Newton or Einstein, fails. The second, less radical interpretation is that the continuation of a constant velocity of rotation out to such large distances means that the galaxy possesses invisible matter beyond its visible boundary.

Physicists and astronomers usually opt for the less radical second interpretation. Thus we have to conclude that a typical galaxy is much more massive than it appears and that its mass, in dark form, extends out to two or three times its visible radius.

Such a finding, even though the less radical of the two, confronts astronomers with the result that the Universe contains far more matter than they can 'see' through their numerous telescopes. How much more matter? Before we consider this question let us look at some more evidence for dark matter on an even grander scale.

Figure 10-7 shows a cluster of galaxies. In a typical cluster one finds hundreds of galaxies moving in one another's gravitational fields. Theory tells us that if this motion has been going on for long enough, the cluster reaches a state of equilibrium when the total kinetic energy of the galaxies and their gravitational potential energy are comparable in magnitude. (See Chapter 2 for a discussion of these energies.) In a typical cluster like that of Figure 10-7 we find that the kinetic energy exceeds the potential energy by a large factor.

So again the astronomers are faced with an unexpected result. And again they rationalize it by arguing that the cluster contains dark matter in such quantities that if we include it in our calculation, the cluster's gravitational potential energy increases to a value at which it becomes comparable to the potential energy.

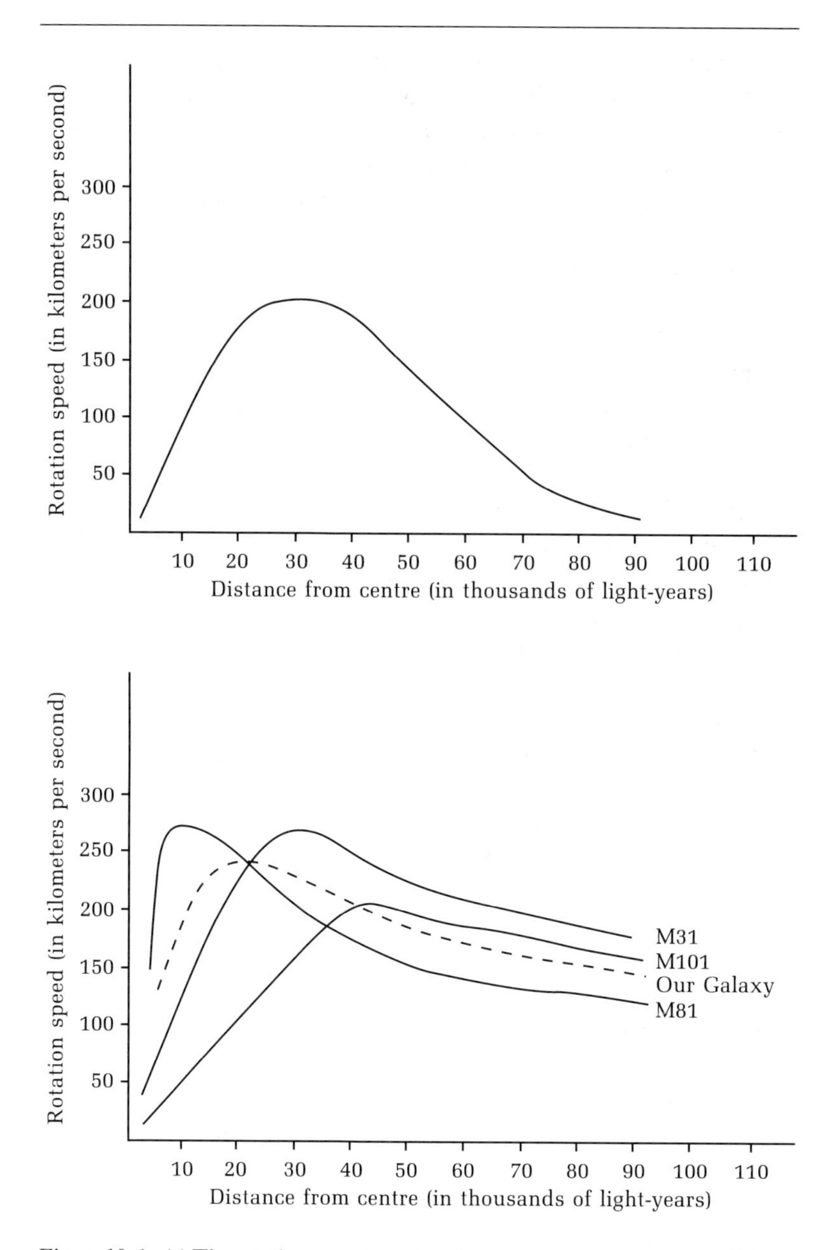


Figure 10-6. (a) The rotation curve based on Newtonian assumptions as per Figure 10.5. (b) The actual rotation curves for a number of galaxies.

Figure 10-7. The Coma cluster of galaxies. (Photograph by courtesy of the National Optical Astronomy Observatories, USA.)

Thus we can say that both in individual galaxies and their clusters a substantial amount of dark matter is present. To the question we raised earlier, 'how much?', the answer differs from one theoretician to another. The ratio of the mass of dark to visible matter in the Universe is certainly greater than 1; it may be easily as high as 10. If we increase it further, then we *can* reach a total density that the supporters of inflation would want. Thus, if the inflationary cosmology is right, then the amount of dark matter is some 30–100 times the amount of visible matter in the Universe.

## WHAT IS DARK MATTER MADE OF?

There are constraints, however, faced by big-bang cosmologists wishing to answer this question. A natural answer would be that the dark matter is made of highly collapsed burnt-out stars that might have become black holes or it is made of nonluminous small lumps of matter not too large compared to a typical planet. The latter possibility, often called a *brown dwarf*, describes an object not more than 10% of the solar mass which is therefore not massive enough to generate its own light through nuclear fusion. A large quantity of such objects could lie in the large haloes of galaxies or in their immediate outer vicinity. They will not be observable by conventional astronomy.

A clever way to detect them was initiated in the early 1990s by two groups of astronomers under the so-called MACHO and EROS programmes*. In these observations the technique of gravitational lensing is used for detecting chunks of dark matter by the effect they produce on a ray of light passing by. The typical scenario is illustrated in Figure 10-8. Imagine a star moving within the Large Magellanic Cloud (LMC), which is a small satellite galaxy to our Milky Way. Suppose a chunk of dark matter comes in the way between ourselves and the star. As the background star moves in the LMC it will move within the Einstein ring of the gravitational lens formed by dark matter (see Chapter 6). At that time the lensing effect will significantly enhance the brightness of the star. The effect, which is the same for light of all colours, will cease when the star has moved away from the Einstein ring. Thus, a temporary brightening of the star will be the indicator of a lens of dark matter.

From 1993 results of this kind started coming in. In contrast to the huge lenses discussed in Chapter 6 for quasars, these are small and hence called *microlenses*! Illustrated in Figure 10-9 (a), (b) can be seen one such event reported by the MACHO programme and another by EROS. It is early days yet to decide whether what is being seen here is a microlensing effect and not an intrinsic light fluctuation of the star due to some internal physical changes. Also, even in the case of microlensing it has been argued by the astronomer Kailash Sahu that the microlensing objects may not be tiny ones with masses less than a tenth of a solar mass but may be ordinary stars in the LMC itself which because of their low mass (less than *half* a solar mass) are very faint.

* MACHO is the acronym for MAssive Compact Halo Object while EROS stands for Expérience de Recherche d'Objets Sombres.

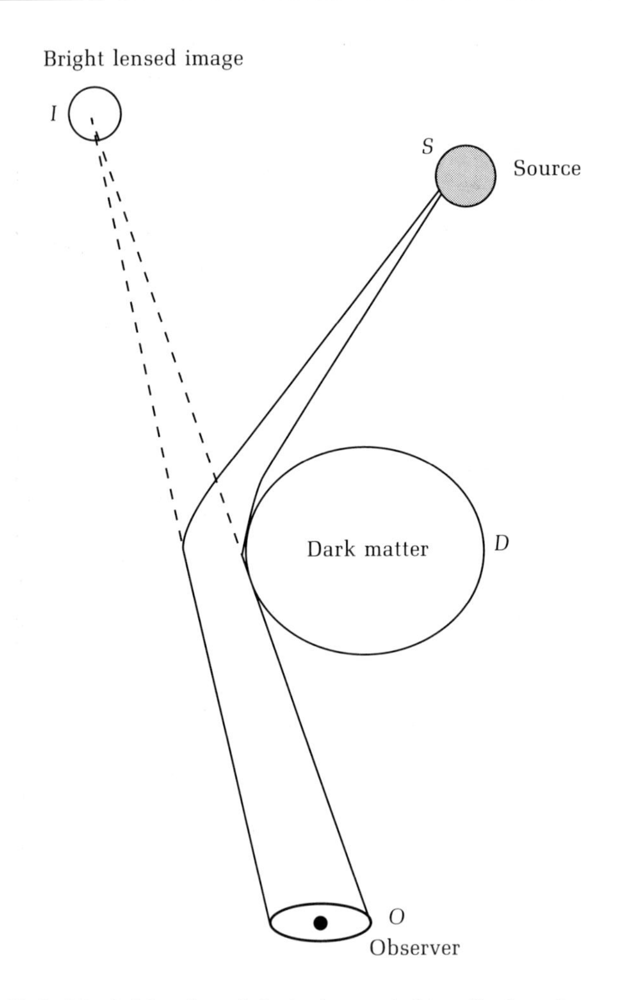


Figure 10-8. The brightening of the background object $S$ takes place as seen by observer $O$ when it is lensed by the intervening lump of dark matter $D$.

Figure 10-9. (a) The brightening produced by gravitational lensing for a MACHO object shows that the rise in intensity lasts for about 34 days. The intensities in blue and red light are separately shown with their comparison at the bottom. The changes in intensity in both colours keep pace which is a signature of gravitational lensing. (Courtesy of the MACHO team and *Nature*: C. Alcock *et al.*, **365**, 621, 1993.) (b) A similar case observed by EROS where the intensity rises and falls over about 27 days. (Courtesy of the EROS team and *Nature*: A. Aubourg *et al.*, **365**, 623, 1993.)

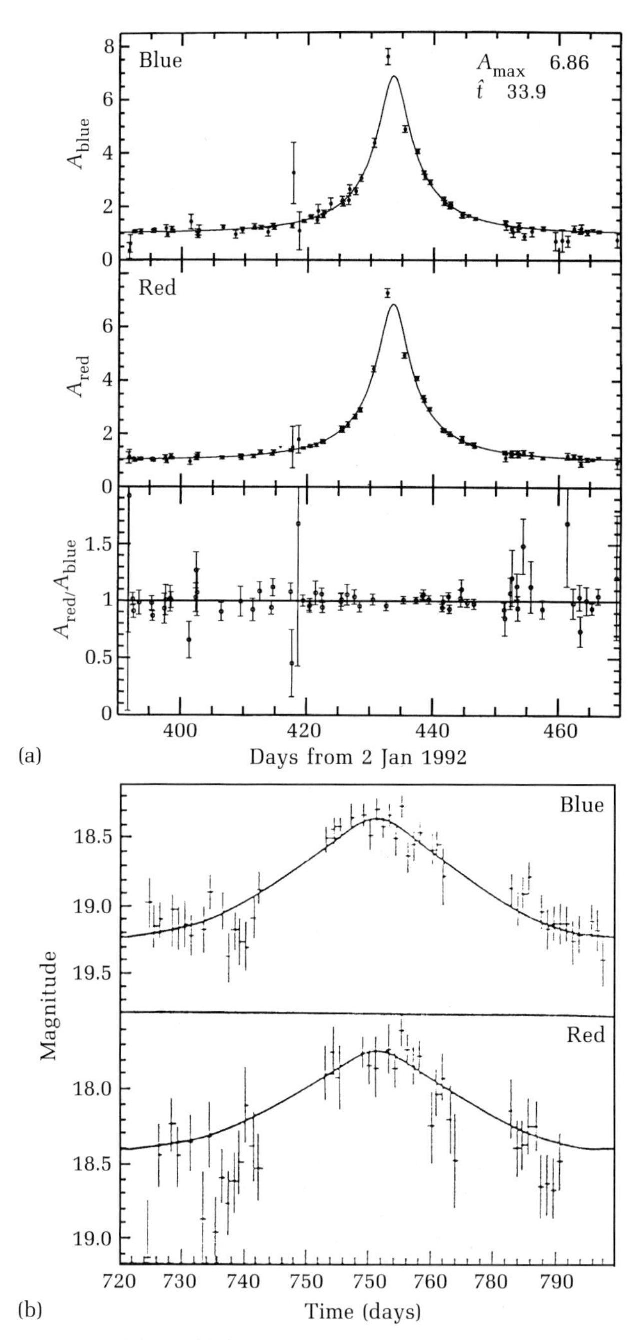


Figure 10-9. For caption see facing page.

The interpretation that all dark matter is of this kind (brown dwarfs, planets, even low-mass stars) will pose serious problem for the big-bang cosmologists, for two reasons.

First, the black holes or brown dwarfs or planets are made of ordinary matter that is found on the Earth: neutrons, protons, and electrons. The heavier of these particles, the proton and the neutron, are collectively called *baryons*. If the density of matter in baryonic form exceeds a certain critical value (which is around $10^{-30}$ grams per cubic centimetre), then a new difficulty arises in hot-big-bang cosmology.

Recall from the previous chapter that the process of primordial nucleosynthesis delivers a tiny fraction of deuterium. This prediction of hot-big-bang cosmology is very sensitively independent on how much baryonic density exists today in the Universe. If it exceeds the above critical value, the deuterium abundance drops sharply to become negligible. So, the big-bang cosmologist would like today's baryonic density of the Universe to stay *below* this critical value. Hence, the dark matter cannot be entirely baryonic.

The second constraint comes from the fluctuations of temperature of the microwave background. Again, recall that at very early times matter and radiation were in continuous interaction till the temperature of the Universe dropped to about 3000 kelvin. Calculations done in the 1960s and the 1970s showed that if this matter was largely baryonic, then it would continue to interact with radiation till this last moment, and the resulting fluctuations of the radiation temperature would be of the order of a few parts in ten thousand. Attempts to look for temperature fluctuations by various experiments failed to show any effect of this order! The fluctuations reported by COBE are too small compared to these theoretical expectations. Thus, if dark matter is present, it has to be largely in a form that does *not* interact with light. This rules out the baryonic option.

So the process of elimination forces us to go for rather esoteric forms of matter that *do not* interact with light. The GUTs and other theories such as supersymmetry from the notebooks of high-energy particle physicists do allow the existence of such particles, although none have been observed in high-energy particle accelerators. Known by names such as 'photino', 'gravitino', 'axion', etc., these could be safely tucked in as dark-matter candidates without violating any observational constraints. Sometimes these particles are called WIMPs, i.e., weakly interacting massive particles.

The literature in this field classifies these nonbaryonic particles under two groups: hot and cold dark matter. A dark-matter candidate is said to belong to this 'hot' category if it was moving very fast (with near-light speed) at the time it ceased to interact with other matter. A neutrino, if it turns out to have low but nonzero mass when at rest, may be an ideal example of hot dark matter. A cold-dark-matter particle on the other hand would have been moving slowly under similar conditions. Typical candidates are photinos, axions, etc.

Although they will take us too far from our discussions on gravity, the cold/hot-dark-matter options have played a major role in current efforts to understand large-scale structure in the Universe. Recipes including cold dark matter or hot dark matter or a mixture of the two are used to supplement ordinary (luminous or visible) matter in order to arrive at a coherent picture of how galaxies and their conglomerations are formed, to understand their large motions, and to explain the kind of imprints on microwave radiation detected by COBE.

The attempts have not yet succeeded. The optimists think that they are bound to succeed one day and that we will soon understand how the large-scale structure came about. The sceptics and cynics may wonder whether this is the correct way to proceed and whether the solution lies in an entirely different direction. In the next chapter we will present an alternative view.

# 11

# *Gravity and the creation of matter*

## WAS THERE REALLY A BIG BANG?

The big-bang cosmology described in the last two chapters has a large following amongst the astronomical community. The models of Friedmann are able to account for the observed expansion of the Universe, for the smooth background of microwave radiation, and for the abundance of light nuclei that cannot be generated inside stars. Are these not good enough reasons for believing in the overall picture?

Playing the devil's advocate in this chapter, let me voice a few dissenting views. First, a scientific theory, howsoever successful it may be, must always be vulnerable to checks of facts and conceptual consistency. Even a well established theory like Newton's had to give way to Einstein's when it was found wanting under these checks (see Chapter 5). The formidable facade of big-bang cosmology is likewise developing cracks that can no longer be plastered over.

The first crack has actually been there right from the beginning and may have been noticed by the reader of Chapter 9. He or she may ask the questions, 'What preceeded the big bang? How did the matter and radiation in the Universe originate in the first place? Does it not contradict the law of conservation of matter and energy?'

These questions cannot be answered within the framework of Einstein's general theory of relativity, which was used to construct the Friedmann models. The moment of 'big bang' is a singular epoch, according to the theory, just as the end of a collapsing object, described in Chapter 7, is in a singularity. Thus, a scientific

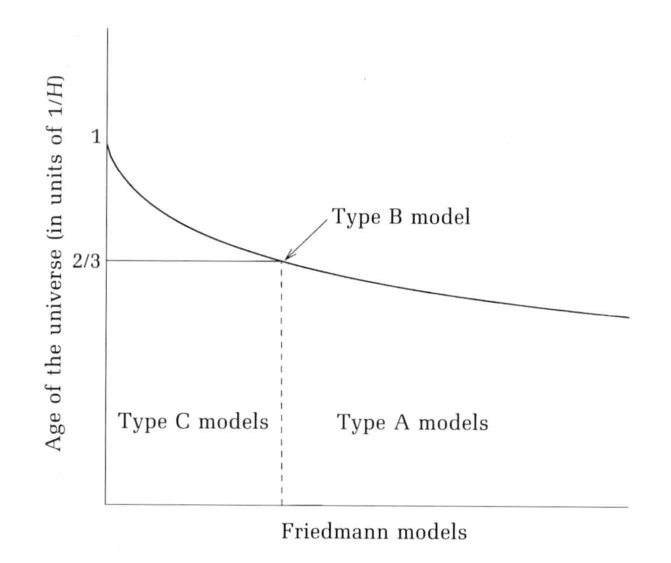


Figure 11-1. The ages of different Friedmann modes in units of $1/H$. Model $B$ has an age two-thirds of $1/H$, while the ages of models of type $A$ range downwards from this value, and of models of type $C$, upwards from this value up to $1/H$.

justification of the big bang is not possible, nor can one talk of what preceded it.

This appearance of a singularity and the contradiction of established rules of science like the law of conservation of matter and energy is serious enough to suggest that at best the theoretical framework describing the big bang is incomplete.

Leaving aside this matter of conceptual inconsistency, let us ask a practical question: 'How old is the Universe?' Going back to the curves of Figure 9-7, we had noted in the discussion following them that $1/H$ represents a time scale. We can now relate it to the question of age of the Universe. Calculations show that the age of the Universe computed for any of the Friedmann models *does not exceed* this time scale. In Figure 11-1 we see how the actual age varies from model to model.

We see here that for model $B$, which is implied by an inflationary phase in the past (Chapter 9), the age is two-thirds of $1/H$. It is higher for $C$-type models (but in any case less than $1/H$) and lower for $A$-type models. We will work with the $B$-type model since it is the one most actively supported by big-bang theoreticians.

Although there has been a long controversy about the 'true' value of Hubble's constant, several measurements now tend to favour a lower value for $1/H$. The measurements in 1994 by observers working on a key project of the Hubble Space Telescope arrived at a value of $1/H$ close to 12.5 billion years as a very reliable estimate with error bars of about 2.5 billion years. Let us take the highest value in the error range, of $12.5 + 2.5 = 15$ billion years for $1/H$, two-thirds of which is 10 billion years. Is this value a high enough value for the age of the Universe?

Hardly! Astronomers over the years have been improving their estimates of ages of stars found in large groups called globular clusters. Several stars in such a cluster can be traced at a stage of evolution where they are just becoming giants (Chapter 4). That is, they have exhausted their hydrogen fuel and are beginning on the next stage of helium fusion. The ages of such stars can be estimated from this information and they come out in the range 12–18 billion years.

A somewhat similar age range turns out to be the answer for the age of our Galaxy from the studies of the decay products of nuclei which undergo very slow radioactivity. This is similar to the techniques the archaeologists use in the carbon dating of old relics by looking at the decay product of excited carbon nuclei. Thus, we find the Universe is not old enough to accommodate all that we see in it.

Evidence turned up by infrared astronomy, on the other hand, has also shown the existence of galaxies that are very young, in which star formation is taking place *now*. It is very hard to incorporate such galaxies in the big-bang scenario, which has all the galaxies forming in a relatively narrow time period several billion years ago. Thus, just as very old objects present an embarrassment to the scenario, so do very young ones.

Even the strongest evidence claimed in support of the hot big bang, the microwave radiation, hides a weakness: its temperature today cannot be predicted by the theory. Thus, the guesses of 5 kelvin by Alpher and Herman (Chapter 10) or of 7 K by Gamow cannot be improved upon by the present theory, which has to accept the measured value of 2.73 K as a given parameter. Thus, the question that remains unanswered is, Why does the microwave background today have a temperature of 2.73 K?

There are other problems when we try to interpret the nature of dark matter. We saw in the previous chapter that, operating under the big-bang constraints, astronomers are forced to think of various esoteric alternatives for the dark matter, alternatives that no one has detected in any terrestrial laboratories or particle accelerators. Indeed, one is reminded of the Hans Andersen story *The Emperor's New Clothes* when theoreticians debate the merits of hot and cold dark matter. Often the claims and counterclaims are made with such conviction that one forgets that none of this has been observed in a laboratory.

Finally, in Chapter 8 we highlighted several events in astrophysics where explosive generation of energy is observed to be taking place. It would have been natural if there were a link between the primordial explosion, the big bang, and these later minibangs. As we discussed there, attempts to interpret such minibangs as delayed explosions after the big bang have failed.

## THE QUASI-STEADY-STATE COSMOLOGY

These objections may not all be admitted by the supporters of the big-bang scenario, but they make up an impressive catalogue to at least prompt the uncommitted to begin thinking of some viable alternatives. Such alternatives are not easy to come by since, like any scientific theory, they too are constrained by facts and requirements of consistency. Since 1993, the author has been associated in one such attempt in collaboration with Fred Hoyle and Geoffrey Burbidge.

Called the *quasi-steady-state cosmology* (QSSC), this theory is a revival of the steady-state cosmology proposed in 1948 by Hermann Bondi, Tommy Gold, and Fred Hoyle. The old steady-state theory talked of a Universe without a beginning and without an end that has been expanding steadily but in which matter is being continually created. The QSSC gives a more realistic version of the old theory as we shall see next.

We begin with the first and the last items in our catalogue of objections to the big-bang. In the QSSC, the basic theory, which is somewhat wider in its framework than general relativity, allows a meaningful discussion of the phenomenon of matter creation, meaningful in the sense that it allows new matter to be created *without violating the law of conservation of matter and energy*. A simple analogy from economics will illustrate how this is done.

Suppose you want to buy a house for which you do not have the capital. So you borrow from a bank or a building society and mortgage the house. You repay the loan plus interest in fixed instalments over a span of say 20 to 25 years. With general inflation the real value of these instalments reduces with time and they become easier to pay off, while the value of your house appreciates. At the end of the mortgage period you have acquired a valuable property whereas you had none to start with! There is one snag though: you need to be above a certain income threshold in order to qualify for the loan.

In the QSSC the role of the bank is played by an energy reservoir called the $C$-field (for *creation* field). Matter is created out of this reservoir. By the law of conservation of matter and energy the $C$-field reservoir will lose the equivalent energy. However, the entire reservoir is of *negative energy*, so that its magnitude increases whenever matter is created at its expense. As the Universe expands, however, this magnitude gets diluted and comes back to its original value. (This is analogous to the bank receiving the loan amount back.) The net effect is that we have matter being created in an expanding Universe!

But why should the Universe expand in such a theory? This happens because in the process of matter creation the $C$-field not only gains extra negative energy, but it also acquires negative stresses. These stresses affect the spacetime geometry exactly as in relativity and they make it expand. Thus, the creation process generates the means of expanding the Universe.

Like the mortgage example, there is a threshold involved here too. For creation to take place, the $C$-field reservoir must acquire enough strength to be able to produce particles of matter. This is not possible everywhere. Only in the vicinity of compact massive objects is this possible. For here the strong gravity in the vicinity of the object increases the strength of the $C$-field to the threshold

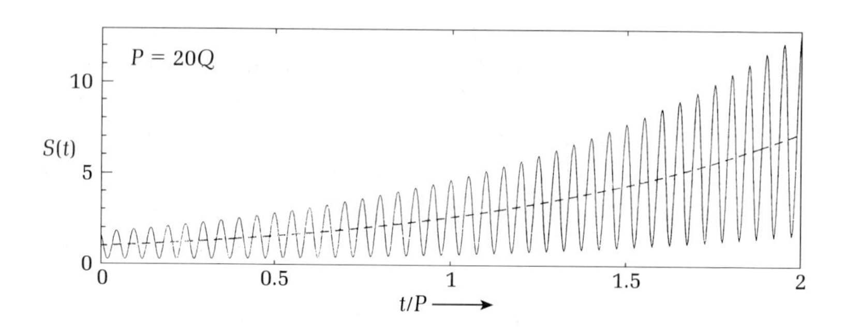


Figure 11-2. The dotted curve describes the long-term trend in the universal expansion while the short-term oscillations are shown by the continuous curve. The time scale ($P$) for the former is 20 times the period ($Q$) for the latter.

level*. Hence creation takes place only near a compact massive object. Not only that, the creation induces local expansion of space which blows the created matter outwards with tremendous energy.

Thus, we have here effectively a mini explosion, a white hole but without a singularity. The objection to white holes that we discussed in Chapter 8, namely that we do not understand their beginning, is thus eliminated. We will consider next how this leads to the expanding Universe.

Imagine several such minibangs or *mini-creation events* going off all over the Universe, each producing a local expansion of space. The combined effect of these is to generate a large-scale expansion of the Universe. If the creation activity were going on steadily, the scale factor $S$ would increase with time according to the so-called exponential function, just as capital grows with time under compound interest. Figure 11-2 shows this behaviour with a dotted curve.

However, the mini-creation events are sensitive to the overall background level through the $C$-field. If the mini-creation events work very efficiently and fast, the Universe will expand very rapidly, faster than the dotted curve of Figure 11-2. A faster expansion brings about a drop in the average background level of the $C$-field.

* Just as a stone dropped into a well picks up kinetic energy, so the $C$-field 'dropped' into the neighbourhood of a compact massive object picks up extra strength owing to the strong gravity of the object.

When this happens, raising the level to the threshold becomes difficult for any compact object and many mini-creation events are switched off. At this stage the expansion slows down to a halt, and with gravity gaining an upper hand over the repulsive forces, it is converted to contraction as in the *A*-type Friedmann models (cf. Chapter 9). With contraction of the Universe the background level of the *C*-field is raised again, thereby reactivating the mini-creation events.

Thus, we have up-and-down creation activity which results in oscillations of the scale factor around the steady value. The continuous undulating curve of Figure 11-2 illustrates this behaviour. Whereas the dotted curve describes a steady-state situation, the continuous curve describes the *quasi-steady state*.

What particles are created in this primary creation process? Theoretical considerations lead to a unique answer, the so-called *Planck particle*, whose mass is of the order

$$\sqrt{\frac{ch}{G}},$$

made of the three fundamental constants $c$ (the speed of light), $h$ (Planck's constant), and $G$ (the constant of gravitation) that we have already encountered. The Planck particle is unstable, however, and quickly decays into more stable particles and radiation. The eventual decay products are the baryons. One Planck particle may produce as many as ten billion billion ($10^{19}$) baryons.

## SOME TESTS AND PREDICTIONS OF THE QSSC

The quasi-steady-state cosmology makes several new predictions that distinguish it from the big-bang cosmology. We highlight a few.

1. A good fit with existing observations like Hubble's law is obtained if we take the longer time scale (dotted curve of Figure 11-2) to be about a million million ($10^{12}$) years and the shorter one, the period of oscillation (continuous curve), to be about 40 billion years. Thus, while the Universe itself is infinitely old, it allows the existence of old as well as

young galaxies. In fact one expects a population of galaxies of varying ages.

2. Galaxies as old as several hundred thousand million years may be quite common. Since stars like the Sun typically complete their evolutionary sequence in times less than this, several galaxies should contain old burnt-out stars which form the dark matter. Thus, all dark matter is baryonic.
3. The formation of light nuclei comes about in a different way to that in hot-big-bang cosmology. These nuclei form from the decay products of the Planck particle, which is the primary form of created matter. This process gives the right abundances of light elements, which are insensitive to the magnitude of baryonic density in the Universe. Thus, there is no constraint (as in the big-bang cosmology) that the baryonic matter density should not exceed a given limit.
4. The microwave background is obtained from the starlight of burnt-out stars from the previous cycles. Recall (from point 2 above) that in a typical cycle of 40 billion years, stars burn out completely. The radiation left behind by them is seen today as the relic microwave radiation. Calculation shows that a temperature very close to 2.7 K is obtained this way. Thus, the temperature is deduced rather than *assumed* as in the big-bang cosmology. The spectrum and small-scale inhomogeneities are exactly like those obtained by COBE (Chapters 10 and 11).
5. Because the Universe alternately expands and contracts, the spectrum of a galaxy suitably placed in the previous cycle (see Figure 11-3) can show a *blueshift*. Such galaxies are very faint, but hopefully with improved technology of the future this prediction may also be tested.
6. The mini-creation events can be powerful sources of gravitational waves, provided they are not isotropic or symmetric about an axis. The type of laser interferometer detectors now under construction should be able to detect a few of them.

This is just a flavour of the new cosmology which is still being developed and tested. Time will tell whether it will survive, die,

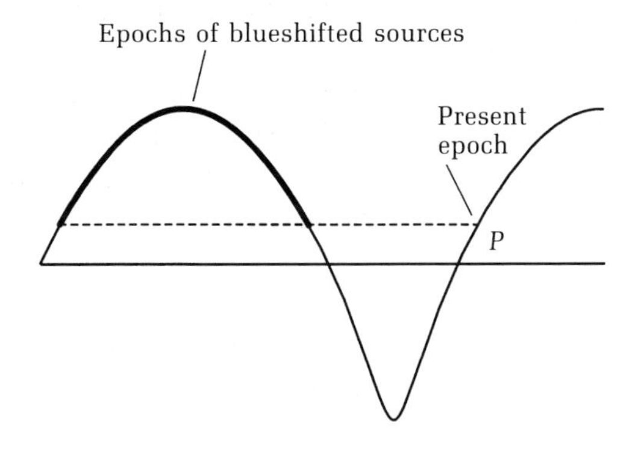


Figure 11-3. To be seen as a blueshifted source of light, the object must have emitted its light (received by us today) when the scale factor of the Universe was *larger* than what it is today. Here the present epoch on the curve of Figure 11-2 (drawn here at an enlarged scale) is shown by point $P$. The thick part of the expansion curve of the previous cycle shows the epochs when the scale factor of the Universe was larger than at $P$. The blueshifted sources belong to these epochs.

or outlive the big-bang cosmology. In the meantime, by offering a serious alternative to the latter, the QSSC may liven up the cosmological scene.

# 12

# *The many faces of gravity*

Our discussion of gravity began with the falling apple and has taken us from ocean tides to the planets, comets, and satellites of the solar system, to the different stages in the evolution of a star, to the curved spacetime of general relativity, to the illusions of gravitational lensing, to the weird effects associated with black holes and white holes, and finally to the large-scale structure of the Universe itself. None of the other basic forces of physics has such a wide range of applications. Although gravity is by far the weakest of the four known basic forces, its effects are the most dramatic.

Indeed, it would be an amusing exercise to speculate on the state of the world if there were no gravity at all! Would atoms and molecules be affected? As far as we know, the presence or absence of gravity does not play a crucial role in the existence and stability of the microworld. The strong, weak, and electromagnetic forces are the main forces at this level. Even at the macroscopic level of the objects we see around us in our daily lives, gravity does not appear to play a crucial role in their constitution or equilibrium. After all, even astronauts have demonstrated that they can live in simulated conditions of weightlessness. Neither the astronauts nor their spacecraft come apart in such circumstances. The basic binding force at this level is the force of electricity and magnetism.

But we can go no further in dispensing with gravity. If we eliminate gravity on a bigger scale, disasters lie in store. With the Earth's gravity gone, there is no force to bind the Earth together as a spherical object, to retain the protective layer of atmosphere around it, or to keep us on its surface. The living systems on the Earth have complex biological systems that have evolved with and have adapted themselves to the explicit presence of the Earth's gravity. Without gravity, it is hard to conjecture how they will survive and evolve.

On a still larger scale, without gravity the Earth would no longer be attracted by the Sun and would take off in a straight path instead of going around the Sun. The Sun itself would no longer be able to maintain stability but would disperse outward. Without gravity, neither the Sun nor any other stars could exist, nor could larger systems such as galaxies and quasars. These are just a few speculations of a zero-gravity Universe. Incomplete as they are, they still give us some indication of the importance of gravity to the physical world.

In spite of its importance and omnipresence, gravity remains shrouded in mystery. Having stated the inverse-square law of gravity, Newton declined to further try to elucidate *why* this law operates. Einstein provided an ingenious connection between gravity and the geometry of spacetime, but even he was conscious of the fact that his description of gravity placed it further apart from the rest of physics. Einstein's prescription eliminated gravity as a force. Instead of simply affecting the motion or equilibrium of a body as any normal physical force is expected to do, gravity transforms the geometry of spacetime around the body. To bridge the gap between gravity and other physical forces, Einstein hoped to construct a *unified field theory* of all physical interactions. He was unsuccessful in this ambitious task, in spite of several years of research. Nevertheless, as we briefly discussed (Chapter 10), today's physicists are pursuing the same goal via a different route. They hope that one day gravity will also come into this fold.

But as of now gravity still remains far apart from this chain of developments, partly because of its unusual description as a geometrical effect of spacetime rather than as a straightforward force. To some extent, the difficulty also lies in identifying what the quantum effects of gravity are. Quantum theory usually relates to the microscopic world. At this scale, gravity is very weak. How can physicists study such effects?

As we increase the speed of a particle, its energy also increases. As a rule, this increases the strength of its interaction with other particles. Physicists use particle accelerators to study the nature of various particle interactions by firing particles at other particles with high energy. We saw in Chapter 10 that the present particle accelerators fall short of the energy required for the operation of GUTs by a whopping factor of a thousand billion. For the study of any quantum effects of gravity and its possible unification with the rest of the forces in physics, the energy needed to be attained is higher by a further factor of several thousands!

These numbers illustrate the difficulties of making progress in understanding the nature of gravity in the terrestrial laboratory. These difficulties in turn imply that, for a long time to come, further understanding of this mysterious interaction must come through astronomy.

We have discussed in this book a few highlights of the astronomical effects of gravity. I now close this discussion by enumerating some unresolved features of these effects.

Over the past decade, considerable research has been done on the physics and astrophysics of black holes. We discussed some of this work in Chapters 7 and 8. Some enthusiasts believe that black holes are the ultimate solution to the energy problem. However, the sceptics remain unconvinced that black holes even exist or have been detected. Defenders of black holes say that the evidence is necessarily circumstantial, and nothing more can be expected. Nevertheless, the note of caution that should accompany any circumstantial evidence is often missing from the claim that 'a black hole exists in such-and-such an object'.

White holes, by comparison, are theoretically directly observable. Many astronomers had serious doubts about their existence; however, as mini-creation events, they can now be placed on a firmer footing.

The Einstein equations lead to a very simple set of models of the Universe, the so-called big-bang models, which were first worked out by Friedmann. These models have been very successful in explaining the redshifts of galaxies, the origin of light nuclei, and the microwave background radiation. Thus, they provide a basic framework in which to tackle the more difficult and detailed issues of cosmology.

However, it is clear that the big bang theory of the Universe leaves the very important question of the origin of the Universe unanswered. Why did the big bang occur? Why, how, and when did matter first appear in the Universe? How did it form into the large-scale structure that we see made of galaxies, clusters, etc.? Why, on the other hand, has the relic radiation remained so smooth? Can we really trust the present laws of physics to such an extent that we can deduce what the Universe was like at the time of the big bang?

Many supporters of the big-bang Universe consider the event of creation to be beyond the domain of science. This belief is reminiscent of Newton's approach. When faced with difficulties of this kind, he postulated 'Divinity' as the solution. Newton's departure from the scientific approach on such occasions pleased the devout but drew protests form his contemporary scientists such as Leibniz.

Questions about the origin of the Universe continue to bother some physicists. Because of these questions, the steady-state theory of the Universe proposed by Bondi, Gold, and Hoyle in 1948 was revised in a more versatile form as the quasi-steady-state cosmology. This theory has the Universe without a beginning and without an end. The matter in this Universe is created not all at once in a single explosion, as in the big-bang Universe, but in mini-creation events around dense massive objects distributed throughout the Universe at all times.

The Universe oscillates on time scales of 40–50 billion years as it expands on a still longer time scale of a thousand billion years. In the 1950s and 1960s the steady-state model had generated considerable controversy and prompted many observational studies of the Universe. Its descendent, the QSSC, is likely to generate a similar reaction, which is all to the good of cosmology and for the subject of gravity; for science eventually benefits from arguments and controversies.

And here we leave our study of the many faces of gravity. Through the study of gravity, nature has so far revealed many of her secrets, but many more are surely being reserved for the future.

# *Index*

Page numbers in italic indicate pages on which the topic is introduced or discussed in greatest detail.